AMORPHOUS SEMICONDUCTORS

Understanding the structural unit of crystalline solids is vital in determining their optical and electronic properties. However, the disordered nature of amorphous semiconductors, where no long-range order is retained, makes it difficult to determine their structure using traditional methods. This book shows how computer modeling can be used to overcome the difficulties that arise in the atomic-scale identification of amorphous semiconductors.

The book explains how to generate a random structure using computer modeling, providing readers with the techniques to construct realistic material structures. It shows how their optical and electronic properties are related to random structures. Readers will be able to understand the characteristic features of disordered semiconductors. The structural and electronic modifications by photon irradiation are also discussed in detail. This book is ideal for both physicists and engineers working in solid state physics, semiconductor engineering, and electrical engineering.

SÁNDOR KUGLER is an Associate Professor at Budapest University of Technology and Economics, Hungary, and Guest Professor at Tokyo Polytechnic University, Japan. He is an overseas editor for the *Japanese Journal of Applied Physics*, and is a well-known expert in amorphous semiconductors and chalcogenide glasses.

KOICHI SHIMAKAWA is a Professor at the University of Pardubice, Czech Republic, a Fellow and Emeritus Professor of Gifu University, Japan, and Senior Researcher at Nagoya Industrial Science Research Institute, Japan. He is an internationally recognized researcher in the field of amorphous solids.

AMORPHOUS SEMICONDUCTORS

SÁNDOR KUGLER
Budapest University of Technology and Economics, Hungary

KOICHI SHIMAKAWA
University of Pardubice, Czech Republic

CAMBRIDGE
UNIVERSITY PRESS

University Printing House, Cambridge CB2 8BS, United Kingdom

Cambridge University Press is part of the University of Cambridge.

It furthers the University's mission by disseminating knowledge in the pursuit of education, learning and research at the highest international levels of excellence.

www.cambridge.org
Information on this title: www.cambridge.org/9781107019348

First published 2015

A catalog record for this publication is available from the British Library

Library of Congress Cataloging in Publication data
Kugler, Sándor, 1950–
Amorphous semiconductors / Sándor Kugler, Budapest University of Technology and Economics, Hungary, Koichi Shimakawa, University of Pardubice, Czech Republic.
pages cm
Includes bibliographical references and subject index.
ISBN 978-1-107-01934-8 (hardback)
1. Amorphous semiconductors. 2. Semiconductors – Computer simulation.
I. Shimakawa, Koichi. II. Title.
QC611.8.A5K84 2015
537.6′223 – dc23 2013022108

ISBN 978-1-107-01934-8 Hardback

To my wife, Szabina, and our children, Szilvia and Zsófia

SK

To my wife, Akiko, and our children, Tetsuro, Shoko, and Ryo

KS

Contents

Preface

Understanding the structural unit of crystalline solids is vital in determining their optical and electronic properties. Determination of the structure of a condensed phase without periodicity is not an easy task. An important objective of the book is to provide an introduction to the reader of how to construct computer modeling of realistic random structures of amorphous IV- and VI-column-element semiconductors and their alloys. Both the merits and drawbacks of the techniques currently used to generate structures using powerful computers are discussed. Furthermore, the structural, electronic, and optical properties of mostly sigma-bonded amorphous semiconductors can be learned.

The basis of this monograph was a course given by Sándor Kugler (SK) during several years at Budapest University of Technology and Economics (and other universities in Europe and Japan), with an extension by Koichi Shimakawa (KS). Our common research experience with amorphous semiconductors extends back more than twenty years. The book is aimed at final-year university students and PhD students in physics, materials science, and chemistry who have already completed introductory courses on quantum mechanics and solid state physics. This book will be useful for both physicists and engineers working in solid state physics, semiconductor engineering, and electrical engineering. For most of the text, no high-level mathematics is needed. This book provides a much wider literature overview than is usual for most handbooks.

A historical overview and a detailed summary of applications are given in the first part of Chapter 1. Readers are informed how to develop further the current technology (photovoltaic cells, thin-film transistors, DVDs, and direct x-ray image detectors for medical use, etc.) using amorphous semiconductors. The rest of the chapter analyzes and answers one of the most exciting questions in the field: what are amorphous semiconductors?

This is followed by a discussion of preparation techniques in Chapter 2. As the glass-transition temperature determines most of the physical properties of glasses, it is briefly discussed, together with Phillip's constraint theory of glassy materials, at the end of the chapter.

The third and longest chapter begins with an important topic, namely how to determine experimentally whether a sample has an amorphous or crystalline phase. The chapter describes atomic-scale computer modeling, including atomic interactions, different simulation methods, and models obtained by structure simulation. The final part of Chapter 3 introduces readers to the most successful commercialized product of chalcogenide glasses, the phase-change materials.

Chapter 4 deals with the electronic behavior of covalently bonded amorphous semiconductors, including defect-free systems and deviation from the ideal networks, i.e. defects. Optical properties of amorphous semiconductors are also described in this chapter.

Chapter 5 presents experimental results of photoinduced changes. It is shown that the structural studies by means of atomic-scale computer simulations are very useful for understanding the experimental results. The photoinduced changes observed in both amorphous chalcogenides and hydrogenated amorphous silicon films are discussed.

We, SK and KS, would like to thank Kazuo Morigaki for introducing us to the physics of amorphous semiconductors. Thanks are also due to Jai Singh (University of Charles Darwin), S.O. Kasap (University of Saskatchewan), Keiji Tanaka (Hokkaido University), Ted Davis (University of Leicester), Stephen Elliott (University of Cambridge), and Takeshi Aoki (Tokyo Polytechnic University) for their powerful discussions. We must also thank Tokyo Polytechnic University (formerly the Tokyo Institute of Polytechnics) for allowing us to use their computing facilities for our large-scale computer simulations.

SK is indebted to many of his colleagues at the Budapest University of Technology and Economics, but he is especially grateful to his four PhD students, Krisztina Kádas (Uppsala University), Krisztián Koháry (University of Exeter), József Hegedüs (University of Helsinki), and Rozália Lukács (Norwegian University of Life Sciences). Furthermore, SK extends his thanks to István László and László Pusztai, with whom he has had the pleasure of working on diverse aspects of research on amorphous semiconductors. Special thanks are also due to Károly Härtlein (Budapest University of Technology and Economics) for his careful work in drawing some of the figures.

KS wishes to thank his colleagues at Gifu University, Shuichi Nonomura and Takashi Itoh, for discussions, and also his PhD student, Yutaka Ikeda (YM systems, Company Executive), who developed a precise measurement system for the photoinduced volume change observed in amorphous chalcogenides. Discussions with Alex Kolobov (Advanced Institute of Industrial Science, Tsukuba), Hiroyoshi Naito (Osaka Prefecture University), Hideo Hosono (Tokyo Institute of Technology), Takashi Uchino (Kobe University), Masaru Aniya (Kumamoto University), and Sergei Baranovskii (University of Marburg) were also very fruitful and enjoyable in promoting our subjects.

Tomas Wagner and Miloslav Frumar (University of Pardubice) provided KS with the opportunity to continue the research work at the Department of General and Inorganic Chemistry (University of Pardubice), after his retirement from Gifu University, supported by grant project CZ.1.07/2.3.00/20/0254 *ReAdMat* financed by the European Union. KS wishes to express his thanks for this opportunity.

1
Introduction

We present a brief historical overview of amorphous semiconductors including their definition.

1.1 Historical overview: science and applications

It is important to include a historical overview of the science and successful applications of amorphous semiconductors in the market in order to understand the current situation. Although science and technology are closely connected, we would venture that some applications have proceeded *without* secure scientific knowledge. This may be a characteristic feature of the field of material science. It is thus a good idea to begin by giving some examples of successful applications.

(1) **Chalcogenides**. Electrophotography (or so-called xerography, a Greek word, meaning "dry writing") has been one of the most successful applications of amorphous selenium (a-Se). The process was demonstrated by C.F. Carlson and O. Kornei in 1938, and modern xerographic processes are the same as those proposed at that time (Pai and Springett, 1993). The Hungarian scientist Pal Selényi first proposed the concept of the photographic process in the 1930s. His pioneering work in electrostatic picture recording formed the basis of xerography. In fact, Selényi published and patented several fundamental ideas of electrography and produced high-quality electrographic copies well before Carlson's proposal (Selényi, 1935a, 1935b, 1936). Films of a-Se have the following unique features: (i) high resistivity (it is a good insulator), and (ii) high photoconductivity. These properties are useful for electronic charging in the dark state and discharging in the photoilluminated state. These technologies

were later applied to laser printers. More recently, a-Se has been replaced by organic polymers for xerography (Weiss and Abkowitz, 2006). The main reason for this is that organic materials cost less.

Most people would recognize the acronym "DVD" (digital versatile disk). The basic operation of the DVD was first proposed by Feinleib *et al.* (1971), although the material $Ge_2Sb_2Te_5$ (also known as GST) has been employed commercially by the Panasonic group in Japan (Yamada *et al.*, 1991). A DVD operates via optically induced phase changes (and hence changes in the reflection coefficient) between amorphous and crystalline states. Using GST, DVDs can be rewritten in excess of one million times, with a crystallization time of less than 50 ns achieved during each rewriting process. The DVD system currently has a memory capacity exceeding 50 GB per disk using a blue semiconductor laser. In the near future, rewritable electrical memory devices will be commercially available that use a phase-change random access memory (PRAM), following the memory switching devices proposed by Ovshinsky (1968). In addition, phase-change materials offer a promising route for the practical realization of new forms of general-purpose and "brain-like" computers that could learn, adapt, and change over time (Wright *et al.*, 2011).

As a-Se is a very sensitive photoconductor, especially for x-rays, due to its high atomic weight, it has been possible to realize a direct x-ray imaging device for use in the medical field. The image of a human hand by Rowlands and Kasap (1997) recalls Wilhelm Röntgen's first x-ray photograph of his wife's hand. This device incorporates a large area of thick (1 mm) a-Se evaporated onto a thin-film transistor (TFT) made from a hydrogenated amorphous silicon (a-Si:H) active matrix array (AMA). The x-ray-induced carriers in a-Se travel along the electric field lines and are collected at their respective biased electrode and storage capacitor. The stored images are then sent directly to the medical specialist's computer. This type of x-ray image sensor is used widely in mammography.

The Japan Broadcasting Corporation, NHK, has utilized the "avalanche photomultiplication" effect in a-Se (Juška, Arlauskas, and Montrimas, 1987; Tanioka, 2007) to create a powerful broadcasting tool. High-gain avalanche rushing amorphous photoconductor (HARP) vidicon tubes have been developed by K. Tanioka and his collaborators, leading to the construction of a HARP vidicon TV camera that is over 100 times more sensitive than a CCD camera.

(2) **Hydrogenated amorphous silicon (a-Si:H)**. In the early 1970s, the oil crisis in the Middle East led to the serious consideration of using photovoltaic (PV) cells as an alternative source of energy. To use PV cells as a viable power

source requires a large area and low cost. Consequently, the first a-Si:H solar cells were fabricated by Carlson and Wronski (1976) at RCA Laboratories in Princeton, NJ. Later, Y. Kuwano's group at Sanyo Co. Ltd. (Japan) was the first to market the PV devices. Common structures comprise p-i-n type heterojunctions. There are several types of PV cells available, the most common being tandem (with dual and triple junctions) a-Si:H configurations, for which more than 10% efficiency is achieved in large-area commercial devices (Carlson *et al.*, 1996). People have recognized the importance of developing PV devices following the Fukushima nuclear power station disaster caused by the earthquake and subsequent tsunami on 11 March 2011.

Thin-film transistors (TFTs) using a-Si:H were first developed in the form of field effect transistors (Powell, 1984; Spear and LeComber, 1984). Two of the most important requirements for TFTs are a high ON/OFF current ratio and a small gate voltage; these are achieved by using a-Si:H. These characteristics mean that TFTs are suitable for use as switching transistors in a liquid crystal display (TFT-LCD), which has completely replaced the former cathode ray tube. Following subsequent improvements in TFT-LCDs, flat-panel displays (FPDs) now produce the clear and large (over 100 cm) images used in TVs and monitors.

(3) **Oxides**. The high demand for flexible and optically transparent TFTs for use in the next-generation FPDs led to the realization of transparent conductive oxides (TCOs) by H. Hosono's group in 1996 (see, for example, Hosono (2006)). Transparent TFTs were developed using ionic oxides such as a-$InGaZnO_4$ (known as a-IGZO). The electron mobility of a-IZGO is larger than that of a-Si:H, and the TFT stability is excellent. Samsung's group in Korea developed the a-IGZO TFT-LCD display that is used commercially in the iPad 3 (Apple Inc.). The high quality and stability of a-IGZO TFT-LCD large-area displays may dominate the "display world" in the near future.

It should be noted that two great discoveries – the amorphous chalcogenides (known as a-chalcogenides or a-Chs) used in electrical memory (Ovshinsky, 1968) and switching devices, and device-quality hydrogenated amorphous silicon (a-Si:H) (Spear and LeComber, 1975) – initiated a vast field of science, which will be briefly reviewed in the following.

The most important scientific issue that resulted from work on amorphous semiconductors was the structural information gained from experiment and theory (in the form of modeling), because the fundamental physical and chemical properties are principally determined by their structure. For tetrahedrally bonded materials such as amorphous germanium (a-Ge) and a-Si, a hand-built random network model was first proposed by Polk (1971), in which 440 atoms and bonds were modeled. Later, computer-generated structures of threefold-coordinated amorphous arsenic (a-As) and twofold-coordinated a-Se were generated by Greaves and Davis (1974) and Long *et al.* (1976), respectively. Molecular dynamic (MD) simulations are now very popular and are used to understand microscopic structures (see, for example, Greaves and Sen (2007)). The average coordination number Z of bonding atoms plays a role in structural properties, and Phillips (1979) proposed a topological constraint model using Z. The magic number $Z_c = 2.4$ (Phillips, 1979) or 2.67 (Tanaka, 1989) may dominate optical and electronic properties in multicomponent glasses. A MD simulation has been also used for the study of photoinduced structural transformation in a-Se (Drabold, Zhang, and Li, 2003; Hegedüs *et al.*, 2005).

A model of the electronic density of states (DOS) for non-crystalline semiconductors was proposed by Cohen, Fritzshe, and Ovshinsky (1969). The DOS separated by a bandgap is not sharp, and extends into the bandgap (where it is known as the band tail). The valence band (VB) originates from the bonding states in a-Si as it does in crystalline Si. However, the VB in a-Chs is formed from a lone-pair band. Amorphous chalcogenides are therefore called lone-pair (LP) semiconductors (Kastner, 1972). In this case, the tailing DOS should be localized, and hence it is known as the "band tail (localized) states." The concept of the mobility edge, which separates the extended and localized states, was then proposed (see, for example, Mott and Davis, 1979; Mott, 1992). It is believed that the localized tail states originate mainly from the distortion of the bond angle, which produces a lack of long-range structural order.

In addition to the localized tail states, there are bonding *defects* that may produce midgap states. In a-Si:H, these defects are identified by Si dangling bonds. These are electronically neutral states. However, in a-Chs, they are believed to be charged dangling bonds, which may be over- or under-coordinated (Street and Mott, 1975; Kastner, Adler, and Fritzshe, 1976).

Optical and electronic transport properties are primarily determined by the electronic DOS. There are no obvious edges to the DOS, and therefore the bandgap is not easy to define. Tauc (1968) defined the bandgap using optical absorption with energy space (without using wave vectors), which has led to it being called the "optical gap" or sometimes the "Tauc gap." In binary and ternary a-Chs, a composition-dependent optical gap was found to be well described using an analogy with mixed crystals (Shimakawa, 1981). Photoluminescence (PL) is dominated by the DOS. Street (1976) initiated work on PL in this field, and recently the technique of PL with wide-lifetime distribution (in the nanosecond to millisecond range) has become well established (Aoki, 2012).

Thermally activated band-type electronic transport occurs in both a-Si:H and amorphous chalcogenides near room temperature. The Meyer–Neldel compensation law for thermally activated processes has been discussed with reference to disordered matter; however, this compensation law is not clearly understood from a theoretical standpoint (Yelon, Movaghar, and Crandall, 2006). At low temperatures, or in defect-dominated materials, transport through localized states is dominant. Mott (1969) developed the variable-range-hopping (VRH) model for a single-phonon carrier hopping process. The importance of small polarons has been pointed out in disordered materials (Emin, 1975), in which multiphonon processes dominate transport. Whether multiphonon or single-phonon processes dominate in amorphous semiconductors is still a matter of debate (Shimakawa and Miyake, 1988; Emin, 2008). Recent results in hydrogenated amorphous/nanocrystalline Si suggest that this factor will depend on temperature (Wienkes, Blackwell, and Kakalios, 2012). Alternating current (ac) transport studies, initiated by Pollak and Gegalle (1961) in crystalline Si, provide information on localized states. Theories involving the continuous time random walk (CTRW) of carriers have developed well, and apply to hopping systems (Dyre and Schröder, 2000).

Time-of-flight measurements developed by Spear and Adams (1966) were applied to crystalline sulfur. In amorphous semiconductors, this technology has been applied to many systems involving a-Si:H and a-Chs to study the drift mobility of both electrons and holes. In most cases, non-Gaussian transport has been discovered. This opened the door for studying dispersive transport in disordered matter (Pfister and Scher, 1978).

The illumination of amorphous semiconductors, effectively with bandgap light, induces various changes on structural and electronic properties. In amorphous chalcogenides, the bandgap decreases with illumination; this is referred to as *photodarkening*, which was first reported by DeNeufville, Moss, and Ovshinsky (1973/1974), and it was confirmed that photodarkening accompanied volume changes (Hamanaka *et al.*, 1976). These effects are of interest in scientific applications, and hence a huge amount of work has been devoted to these topics (see the reviews by Tanaka (1990), Pfeiffer, Paesler, and Agarwal (1991), and Shimakawa, Kolobov, and Elliott (1995)). Photoinduced volume changes in a-Se films have been discussed using MD simulations (Hegedüs *et al.*, 2005). Giant photoexpansion (Hisakuni and Tanaka, 1995) and photodeformations (Tanaka and Mikami, 2009) have produced the interesting terms *photofluidity* and *optical force*, respectively.

A decrease in the photoconductivity of a-Si:H after photoillumination was first discovered by Staebler and Wronski (1977), and is called the Staebler–Wronski (SW) effect. This effect is problematic for applications and hence it is also called *photodegradation*. Photodegradation is attributed to defect creation, and understanding it is a very important subject (see, for example, Street (1991), Redfield and Bube (1996), Morigaki (1999), and Singh and Shimakawa (2003)). Similar effects on photoconductivity have also been found in amorphous chalcogenides (Shimakawa, 1986).

1.2 Definitions

Crystalline and non-crystalline structure

For the reader unfamiliar with amorphous semiconductors, we present here a brief explanation of the most essential definitions. Firstly, we must stress that there is a considerable amount of confusion in the scientific literature concerning the terms non-crystalline, amorphous, glassy, vitreous, randomness, disorder, liquid, and even crystalline. A first important question is whether an atomic structure is crystalline or non-crystalline (amorphous). A perfect crystal is one in which the atoms, or a group of atoms, are arranged periodically in three dimensions to an infinite extent and are rigidly fixed at their thermal equilibrium. This mathematical model of atomic configurations provides us with a relatively easy method for calculating the different properties of condensed matter, which may be found in books on theoretical solid state physics.

A more realistic arrangement is an imperfect crystal in which the atoms form a pattern that repeats periodically only to a finite extent. In fact, this may seem more realistic, but we still have a serious problem in that this type of imperfect crystal behaves counter to thermodynamics. This is because, at non-zero temperatures, defects form in any atomic configuration when the crystals are in their equilibrium states. Real crystals are, of course, not only finite in size, but also contain imperfections such as vacancies, interstitial (foreign or self) atoms, dislocations, impurities, distortions associated with the surface, etc. Furthermore, another effect must be taken into account at finite temperature. The random motion of atoms at their equilibrium positions also weakly destroys the perfect periodicity at any given moment in time. These defects cause distortions in the crystal lattice, but we do not consider these real crystals to be amorphous solids. Translational symmetry more or less remains.

Until 1992, a crystal was defined by the International Union of Crystallography as "a substance, in which the constituent atoms, molecules, or ions are packed in a regularly ordered, repeating three-dimensional pattern." In 1984 (two years after their discovery), Shechtman *et al.* (1984) published a paper on rapidly solidified alloys of aluminum with 10–14% manganese that possess icosahedral symmetry in combination with long-range order, named *quasicrystals*, in clear violation of the above definition of a crystal. Since this discovery, hundreds of similar atomic structures have been synthesized in laboratories around the world. Recently, the naturally occurring quasicrystalline mineral icosahedrite has been identified in a sample from the Khatyrka river in Chukotskii Autonomous Okrug, Far-Eastern Region, Russia. The International Union of Crystallography has had to modify their declaration, and the new and broader definition of crystal became "any solid having an essentially discrete diffraction diagram."

Other types of deviation from perfection occur in crystals. Consider a perfect lattice, but one in which each atomic site possesses a randomly oriented spin or other magnetic moment. Another partially ordered arrangement is possible in materials built up of large near-spherical molecules. Molecular centers of mass can be found in crystalline sites where the directions of molecular axes are randomly distributed. A typical example of the so-called *plastic crystals* is the solid crystalline phase of C_{60} molecules. A similar example occurs when a binary (ternary) alloy has two (three) different, randomly distributed atoms inside a perfect lattice. Are they crystalline or non-crystalline? This issue does not occur for amorphous materials.

A *liquid crystal* is a state of matter that has properties between those of a conventional liquid and those of a solid crystal. Rigid rod-shaped or disc-like molecules develop in a well-defined direction. The centers of mass of the molecules are distributed randomly in two or three dimensions. How does one distinguish between the condensed phase and the liquid phase? A solid is a phase whose shear viscosity exceeds $10^{13.6}$ Pa s (N s m^{-2}), although this threshold is rather arbitrary. Values for the viscosity of common liquids at room temperature are on the order of 10^{-3} Pa s. For some examples, see Table 1.1. Liquid crystals are anisotropic fluids.

A *granular material* is a collection of distinct macroscopic particles with randomly distributed centers of mass, such as sand or peanuts in a container. There is a weak interaction between macroscopic particles. Their behavior is

Table 1.1. *Shear viscosity values for some materials.*

Substance	Temperature (°C)	Shear viscosity (Pa s)
Water	90	0.31
Water	10	1.31
Honey	room temperature	2–10
Bitumen	20	10^8
Ice	–13	15×10^{12}
Window glass	room temperature	10^{40}

differently from that of solids, liquids, and gases, and this has led many to characterize granular materials as a new form of matter. A similar topology can be observed in the "microworld." Polycrystalline materials (poly is a Greek word meaning "many") consist solely of crystalline grains, separated by grain boundaries, and the interaction between them is strong. Furthermore, in some cases crystalline grains do not touch each other; rather, they are embedded into a disordered atomic environment. The grain size determines whether these materials are named *nanocrystalline* or *microcrystalline* materials.

Amorphous, glassy materials

After our brief overview, we can provide the following definition for amorphous materials: amorphous materials exist in the condensed phase and do not possess the long-range translational order (periodicity) of atomic structures. By "long range" we mean over 10 nm. We may also talk about short-range order (also called "local order," below 1 nm) and medium-range order (in the range 1–10 nm). Following the discovery of quasicrystals, another sentence must be added to the previous definition of amorphous materials.

Amorphous materials exist in the condensed phase and do not possess the long-range translational order (periodicity) of atomic structures. They do not exhibit a discrete diffraction pattern.

Using this terminology, the set of amorphous materials has within it a fundamental subset *glasses*, where a glass is an amorphous solid that exhibits a glass transition. If a liquid (melt) is cooled very rapidly so that crystallization can be bypassed, the disordered structure can be frozen-in. This disordered condensed phase is known as a glass. Such a glass-forming process involves the supercooling of a liquid below its normal freezing point. The freezing of a crystalline solid is a first-order thermodynamic phase transition; there are discontinuities in first-order thermodynamic variables such as entropy, $S = -(dG/dT)_p$, or volume, $V = (dG/dp)_T$, at the transition (where G is the Gibbs free energy, $G(T, p, N) = E - TS + pV$). The transformation from a melt to the glassy phase is a transition in which there is no discontinuity in first-order thermodynamic variables at the glass-transition temperature, T_g. However, in second-order thermodynamic variables, such as the calorimetric heat capacity at constant pressure, $C_p = T(dS/dT)_p = -T(d^2G/dT^2)_p$, a discontinuity can be observed.

Groups IV and VI elements and their alloys

For us, the most interesting class of materials comprises the amorphous semiconductors. An important task therefore is to define a semiconductor. A simple definition would be that a semiconductor belongs to a class of materials that have an energy gap of 1–2 eV in their electronic DOS and whose resistivity values fall between that of an insulator and a good conductor. A more accurate definition of crystalline semiconductors is associated with the temperature dependence of resistivity:

$$\rho(T) = \rho_0 \exp(\varepsilon_0 / k_B T), \tag{1.1}$$

where ρ_0 and ε_0 are constants, k_B is the Boltzmann factor, and T is the temperature. Note that ρ decreases with increasing T, which is opposite to the case for conventional metals (where ρ is proportional to T). Most chalcogenide glasses and hydrogenated a-Si follow the above equation, whereas pure a-Si and a-Ge do not.

Atomic structure is one of the main features that distinguish between different electronic transport properties. Carbon provides us with a simple example. The diamond crystal arrangement of carbon atoms creates a good insulator, with about a 5.5 eV gap in the electronic DOS. However, the non-crystalline configuration of carbon displays semiconductor properties, and the graphitic atomic structure is a conductor. Furthermore, superconductivity at 30 K in caesium-doped C_{60} has been observed.

Pure amorphous semiconductor materials are located in the even-numbered groups of the periodic table, such as group IV (carbon, silicon, and germanium) and group VI (selenium and tellurium). In group VI, stable allotropes of sulfur are excellent electrical insulators.

The basis of this structure was formulated by Sir Nevill Francis Mott (Mott, 1969), who concluded via his famous $(8-N)$ rule that there are no dangling bonds inside such a material. This is explained as follows. In a glass, any atom appears in such a way that it retains its natural coordination. The number of covalent bonds $Z = 8 - N$, where N is the number of valence electrons. (We consider elements only in groups IV, V, and VI.) Further, an additional rule states that $Z = N$ if $N < 4$. The most important consequence of this was that amorphous semiconductors *cannot* be doped. Today, however, we know this is not the case.

References

Aoki, T. (2012). Photoluminescence spectroscopy. In *Characterization of Materials*, 2nd edn., ed. E. N. Kaufmann. Hoboken, NJ: John Wiley and Sons, pp. 1158–1169.

Carlson, D.E. and Wronski, C.R. (1976). Amorphous silicon solar cell. *Appl. Phys. Lett.*, **28**, 671–673.

Carlson, D.E., Arya, R.R., Bennett, M. *et al.* (1996). Commercialization of multijunction amorphous silicon modules, *Conf. Record of the Twenty Fifth IEEE, Photovoltaic Specialists Conf.*, Crystal City, Washington D.C., May 13–17, 1996, pp. 1023–1028.

Cohen, M.H., Fritzshe, H., and Ovshinsky, S.R. (1969). Simple band model for amorphous semiconducting alloys. *Phys. Rev. Lett.*, **22**, 1065–1068.

DeNeufville, J.P., Moss, S.C., and Ovshinsky, S.R. (1973/1974). Photostructural transformations in amorphous As_2Se_3 and As_2S_3 films. *J. Non-Cryst. Solids*, **13**, 191–223.

Drabold, D.A., Zhang, X., and Li, J. (2003). First-principles molecular dynamics and photostructural response in amorphous silicon and chalcogenide glasses. In *Photo-Induced Metastability in Amorphous Semiconductors*, ed. A.V. Kolobov. Weinheim: Wiley-VCH, pp. 260–278.

Dyre, J.C. and Schröder, T.B. (2000). Universality of AC conduction in disordered solids. *Rev. Mod. Phys.*, **72**, 873–892.

Emin, D. (1975). Phonon-assisted transition rate: optical-phonon-assisted hopping in solids. *Adv. Phys.*, **24**, 135–218.

Emin, D. (2008). Generalized adiabatic polaron hopping: Meyer-Neldel compensation and Poole-Frenkel behaviour. *Phys. Rev. Lett.*, **100**, 16602–16604.

Feinleib, J., deNeufville, J., Moss, S.C., and Ovshinsky, S.R. (1971). Rapid reversible light-induced crystallization of amorphous semiconductors. *Appl. Phys. Lett.*, **18**, 254–257.

Greaves, G.N. and Davis, E.A. (1974). A continuous random network model with three-fold coordination. *Philos. Mag.*, **29**, 1201–1206.

Greaves, G.N. and Sen, S. (2007). On organic glasses, glass-forming liquids and amorphizing solids. *Adv. Phys.*, **56**, 1–166.

Hamanaka, H., Tanaka, K., Matsuda, A., and Iizima, S. (1976). Reversible photo-induced volume changes in evaporated As_2S_3 and $As_4Se_5Ge_1$ films. *Solid State Commun.*, **19**, 499–501.

Hegedüs, J., Kohary, K., Pettifor, D.G., Shimakawa, K., and Kugler, S. (2005). Photo-induced volume changes in amorphous selenium. *Phys. Rev. Lett.*, **95**, 206803, 1–4.

Hisakuni, H. and Tanaka, K. (1995). Optical microfabrication of chalcogenide glasses. *Science*, **270**, 217–218.

Hosono, H. (2006). Ionic amorphous oxide semiconductors: material design, carrier transport, and device application. *J. Non-Cryst. Solids*, **352**, 851–858.

Juška, G., Arlauskas, K., and Montrimas, E. (1987). Features of carriers at very high electric fields in a-Se and a-Si:H. *J. Non-Cryst. Solids*, **97–98**, 559–561.

Kastner, M.A. (1972). Bonding bands, lone-pair bands, and impurity states in chalcogenide semiconductors. *Phys. Rev. Lett.*, **28**, 355–357.

Kastner, M.A., Adler, D., and Fritzshe, H. (1976). Valence-alternation model for localized gap states in lone-pair semiconductors. *Phys. Rev. Lett.*, **37**, 1504–1507.

Long, M., Galison, P., Alben, R., and Connell, G.A.N. (1976). Model for structure of amorphous selenium and tellurium. *Phys. Rev. B*, **13**, 1821–1829.

Morigaki, K. (1999). *Physics of Amorphous Semiconductors*. London: Imperial College Press and World Scientific.

Mott, N.F. (1969). Conduction in non-crystalline materials. *Phil. Mag.*, **19**, 835–852.

Mott, N.F. (1992). *Conduction in Non-Crystalline Materials*, 2nd edn. Oxford: Clarendon Press.

Mott, N.F. and Davis, E.A. (1979). *Electronic Processes in Non-Crystalline Materials*, 2nd edn. Oxford: Clarendon Press.

Ovshinsky, S.R. (1968). Reversible electrical switching phenomena in disordered structures. *Phys. Rev. Lett.*, **21**, 1450–1453.

Pai, D.M. and Springett, B.E. (1993). Physics of electrophotography. *Rev. Mod. Phys.*, **65**, 163–211.

Pfeiffer, G., Paesler, M.A., and Agarwal, S.C. (1991). Reversible photodarkening of amorphous arsenic chalcogens. *J. Non-Cryst. Solids*, **130**, 111–143.

Pfister, G. and Scher, H. (1978). Dispersive (non-Gaussian) transient transport in disordered solids. *Adv. Phys.*, **27**, 747–798.

Phillips, J.C. (1979). Topology of covalent non-crystalline solids I: short-range order in chalcogenide alloys, *J. Non-Cryst. Solids*, **34**, 153–181.

Polk, D.E. (1971). Structural model for amorphous silicon and germanium. *J. Non-Cryst. Solids*, **5**, 365–376.

Pollak, M. and Gegalle, T.H. (1961). Low-frequency conductivity due to hopping processes in silicon. *Phys. Rev.*, **122**, 1742–1753.

Powell, M.J. (1984). Material properties controlling the performance of amorphous silicon thin film transistors. *MRS Symp. Proc.*, **33**, 259–274.

Rowlands, J.A. and Kasap, S.O. (1997). Amorphous semiconductors usher in digital X-ray imaging. *Phys. Today*, **50**, 24–31.

Redfield, D. and Bube, R.H. (1996). *Photoinduced Defects in Semiconductors*. Cambridge: Cambridge University Press.

Selényi, P. (1935a). Elektrographie, ein neues elektrostatisches Aufzeichnungsverfahren und seine Anwendungen. *Elektrotech. Z.*, **56**(35), 961–963.

Selényi, P. (1935b). Methoden, Ergebnisse und Aussichten des elektrostatischen Aufzeichnungsverfahrens. (Elektrographie). *Z. Tech. Phys.*, **12**, 607–614.

Selényi, P. (1936). Electrostatic recording (electronography). *Electronics*, Apr., 44–46.

Shechtman, D., Blech, I., Gratias, D., and Cahn, J.W. (1984). Metallic phase with long-range orientational order and no translational symmetry. *Phys. Rev. Lett.*, **53**, 1951–1953.

Shimakawa, K. (1981). On the compositional dependence of the optical gap in amorphous semiconducting alloys. *J. Non-Cryst. Solids*, **43**, 229–244.

Shimakawa, K. (1986). Persistent photocurrent in amorphous chalcogenides. *Phys. Rev. B*, **34**, 8703–8708.

Shimakawa, K. and Miyake, K. (1988). Multiphonon tunnelling conduction of localized π electrons in amorphous carbon films. *Phys. Rev. Lett.*, **61**, 994–996.

Shimakawa, K., Kolobov, A.V., and Elliott, S.R. (1995). Photoinduced effects and metastability in amorphous semiconductors and insulators. *Adv. Phys.*, **44**, 475–588.

Singh, J. and Shimakawa, K. (2003). *Advances in Amorphous Semiconductors*. London and New York: CRC Press.

Spear, W.E. and Adams, A.R. (1966). Photogeneration of charge carriers and related optical properties in orthorhombic sulphur. *J. Phys. Chem. Solids*, **27**, 281–290.

Spear, W.E. and LeComber, P.G. (1975). Substitutional doping of amorphous silicon. *Solid State Commun.*, **17**, 1193–1196.

Spear, W.E. and LeComber, P.G. (1984). The development of the a-Si: H field-effect transistor and its possible applications. *Semicond. Semimetals*, **21D**, 89–114.

Staebler, D.L. and Wronski, C.R. (1977). Amorphous silicon solar cell. *Appl. Phys. Lett.*, **28**, 671–673.

Street, R.A. (1976). Luminescence in amorphous semiconductors. *Adv. Phys.*, **25**, 397–453.

Street, R.A. (1991). *Hydrogenated Amorphous Silicon*. Cambridge: Cambridge University Press.

Street, R.A. and Mott, N.F. (1975). States in the gap in glassy semiconductors. *Phys. Rev. Lett.*, **35**, 1293–1296.

Tanaka, K. (1989). Structural phase transitions in chalcogenide glasses. *Phys. Rev. B*, **39**, 1270–1279.

Tanaka, K. (1990). Photoinduced structural changes in chalcogenide glasses. *Rev. Solid State Sci.*, **4**, 641–659.

Tanaka, K. and Mikami, M. (2009). Photo-induced deformations in chalcogenide glass. *J. Opt. Adv. Mater.*, **11**, 1885–1890.

Tanioka, K. (2007). The ultra sensitive TV pickup tube from conception to recent development. *J. Mater. Sci.*, **18**, S321–S325.

Tauc, J. (1968). Optical properties and electronic structure of amorphous Ge and Si. *Mater. Res. Bull.*, **3**, 37–46.

Weiss, D.S. and Abkowitz, M. (2006). Organic photoconductors. In *Springer Handbook of Electronic and Photonic Materials*, eds. S.O. Kasap and P. Capper. New York: Springer, chap. 39.

Wienkes, L.R., Blackwell, C., and Kakalios, J. (2012). Electronic transport in doped mixed-phase hydrogenated amorphous/nanocrystalline silicon thin films. *Appl. Phys. Lett.*, **100**, 072105, 1–3.

Wright, C.D., Yanwei, L., Kohary, K.I., Aziz, M.A., and Hicken, R.J. (2011). Arithmetic and biologically-inspired computing using phase-change materials. *Adv. Mater.*, **23**, 3408–3413.

Yamada, N., Ohno, E., Nishiuchi, K., Akahira, N., and Takao, M. (1991). Rapid phase-transitions of GeTe-Sb_2Te_3 pseudobinary amorphous thin films for an optical disk memory. *J. Appl. Phys.*, **69**, 2849–2856.

Yelon, A., Movaghar, B., and Crandall, R.S. (2006). Multi-excitation entropy: its role in thermodynamics and kinetics. *Rep. Prog. Phys.*, **69**, 1145–1194.

2
Preparation techniques

Amorphous semiconductors are prepared via non-equilibrium processes. Many preparation techniques are possible, depending on what kinds of materials are required for research and/or application. Quenching from the liquid state (otherwise known as melt-quenching) is a popular technique used for so-called glasses and glass fibers. When thin films are required, techniques such as evaporation, sputtering, and chemical vapor deposition (CVD) are adopted. Ion bombardment and high-intensity light directed into crystalline solids also produces amorphous materials. In this chapter, we discuss various methods used in the preparation of amorphous semiconductors. As the nature of glass and amorphous semiconductors is closely related to the structural constraints of the materials, the Phillips constraint theory will also be discussed.

2.1 Growth of thin-film forms

Amorphous Si, Ge, C, and related compounds are prepared via condensation from the gas phase, which usually produces thin-film forms. We briefly review in this section the most popular preparation techniques for the growth of thin-film forms of amorphous semiconductors. These techniques can be roughly classified into two categories: physical and chemical depositions of thin films. *Thermal evaporation* and *sputtering* are physical vapor deposition (PVD) techniques, and the well-known glow-discharge technique comes under chemical vapor deposition. We can clearly distinguish between CVD and PVD techniques. In CVD, there is no target (source material) present: the gas is introduced and decomposed in some way, and these decomposed materials are then deposited onto a substrate as a thin film. The CVD techniques discussed here include "glow-discharge," which is called now *plasma-enhanced chemical vapor deposition* (PECVD), and *hot-wire chemical vapor deposition* (HWCVD), which uses thermal energy. We discuss all of these techniques later, but let us start with PVD.

Thermal evaporation

This is the most conventional method used in preparing amorphous materials. The starting materials comprise ingots or powders, which are mounted in a sample boat and melted in a vacuum chamber at around 10^{-6} Torr ($\sim 10^{-4}$ Pa). To achieve this low pressure, an oil diffusion pump is used in conjunction with a liquid-nitrogen-cooled trap. Note that the boat itself acts as the electrical heater. High-melting-point metal boats, made for example from molybdenum (Mo) or tungsten (W), are normally used. This technique is suitable for low-melting-point materials, and the vapor is condensed at arbitrary temperatures onto the desired substrates. The deposition rate, which is usually about 10 nm s^{-1}, is controllable and depends on the boat temperature. The thickness of the film can be controlled by the deposition time. Well-known amorphous chalcogenides, for example Se, $As_2Se(S)_3$, $GeSe(S)_2$, etc., can be prepared using this technique. The main disadvantage of thermal evaporation is in ensuring that the composition of the resulting films is the same as that of the starting materials.

Device-quality a-Se thick (>1 mm!) films, which are commercially utilized for direct x-ray image detectors (Rowlands and Kasap, 1997) and high-gain avalanche rushing amorphous photoconductor (HARP) vidicon TV cameras (Tanioka, 2007), are prepared by the traditional thermal evaporation technique.

Electron-beam and arc evaporation

Materials with higher melting points, for example, Si, C, and SiC, are not easy to vaporize, and hence a simple evaporation technique is not useful for these materials. Instead of heating the substances in boats, as in thermal evaporation, high-energy electron bombardment (from an electron gun placed in the vacuum chamber) is employed to vaporize such high-melting-point materials.

Arc discharge is also used as a heat source for electrically conducting materials, such as graphitic carbons (Shimakawa *et al.*, 1991). For example, a high current is passed through two pieces of graphite, causing the area to heat up. This process is carried out either in a vacuum or in an atmosphere of Ar or He. (In He, the deposited films contain soccer-ball-shaped C_{60} molecules or carbon nanotubes.) The deposition rate for this method is around 10 nm s^{-1}.

Sputtering

The usual sputtering process comprises a radio-frequency (rf, 13 MHz) high-electric field applied between the target (source material) and the substrate electrodes. A sputtering gas (0.13–2.7 Pa) (for example, Ar) is introduced into a chamber and is pumped away by a vacuum system. The substrate temperature is easy to control. Positively ionized gases produced by the rf field supply kinetic energy to the atoms on the surface of the source material target. Atoms or molecules dissociated from the target are deposited onto the substrate in the form of films. Reactive gases such as H_2 and/or N_2 are also used for sputtering, in a technique known as *reactive sputtering*, in which the gases are incorporated into the deposited films (for example, a-Si:H, a-C:H, a-$Si_{1-x}N_x$, etc.). The deposition rate is lower than that for evaporation techniques, i.e. less than 1 nm s^{-1}, although this will depend greatly on the sputtering conditions.

The sputtering process is more complicated than thermal evaporation; for example, we must pay attention to the sputtering gas pressure, the rf power applied to the target, the bias voltage of the target or substrate, etc. However, sputtering is superior to evaporation for multicomponent systems, and the composition of the films is almost the same as that of the source materials (targets) because the sputtering rates for different elements are on the same order.

When a magnetic field is applied perpendicular to the electrodes, deposition is enhanced; this is because the ionized gases are confined in a spiral motion by the magnetic field (Lorentz force). This method is called *magnetron sputtering*, and it significantly improves the deposition rate. Magnetron sputtering is therefore suitable and cost-effective for large-scale production, and is used for optically transparent conductive oxides (TCOs) such as indium tin oxide (ITO) and semiconducting IGZO for the flat-panel display industry (Kamiya, Nomura, and Hosono, 2009). Note that device-quality a-IZGO films are also prepared using pulsed laser deposition, which will be described later. So called phase-change materials, such as the chalcogenide $Ge_2Sb_2Te_5$ used in optical memory devices (such as DVDs), are also prepared using the magnetron sputtering technique.

Plasma-enhanced chemical vapor deposition

This technique was previously known as glow-discharge deposition, whereby hydrogenated amorphous silicon (a-Si:H) films were deposited by means of the decomposition of SiH_4 gas with the help of glow-discharge (Spear and LeComber, 1975). It is now called plasma-enhanced chemical vapor deposition

(PECVD) (Schiff, Hegedus, and Deng, 2010). The plasma is introduced via the application of a rf field (usually 13.6 MHz), and is therefore referred to as rf PECVD. To prepare n or p type Si:H films, PH_3 or B_2H_6 gas is introduced onto SiH_4. Radio-frequency PECVD is a very popular technique and provides high-quality uniform films, although the deposition rate can be as low as 0.3 nm s^{-1}. The development of this technique, by controlling the gas pressure (0.05–2 Torr), rf power (10–100 mW cm^{-2}), and substrate temperature (150–350 °C) etc., has led to growing numbers of commercial applications, such as large-area photovoltaics and thin-film transistors.

Radio-frequency PECVD is also widely used in the preparation of hydrogenated amorphous carbon (a-C:H) thin films and nanoparticles. Hydrocarbons, such as CH_4 or C_6H_6, are the main precursor materials. Changing the pressure and the rf power allows the preparation of a-C films with remarkably different characters, ranging from soft and porous polymer-like a-C:H and hard diamond-like carbon layers to highly conductive graphitic structures (Veres, Tóth, and Koós, 2008). At low ion energies, intact molecules of the feed gas were found to be incorporated into the matrix of the amorphous films (Veres, Koós, and Pócsik, 2002). Hydrogenated amorphous carbon nanoparticles are prepared by using high gas pressures together with low ion energies and a benzene precursor (so-called "dusty plasma") (Pócsik *et al.*, 2003). Specific dusty-plasma conditions allow amorphous nanoparticles and thin films to be obtained simultaneously (Veres *et al.*, 2005).

When PECVD is carried out at very high frequencies (40–100 MHz) it is called VHF PECVD; this technique improves the deposition rate, which can reach 2 nm s^{-1}. VHF PECVD also produces high-quality films, although poor uniformity is a problem. Microwave PECVD performed at 2.45 GHz (MW PECVD) significantly enhances the deposition rate (up to 10 nm s^{-1}), although device-quality a-Si:H films are not easy to obtain. Note, however, that VHF PECVD and MW PECVD are very useful in the preparation of microcrystalline Si films (μc-Si:H), as the deposition rate of VHF PECVD for μc-Si:H is very small. A high deposition rate for low-cost mass production can be a necessary condition.

Hot-wire chemical vapor deposition

If we are interested in achieving a very high deposition rate in a-Si:H and related films, hot-wire chemical deposition (HWCVD) is a promising technique, with deposition rates reaching up to 15–30 nm s^{-1} (Mahan *et al.*, 2001). The set-up for a HWCVD system is similar to that for rf PECVD, except that the rf electrode is replaced by a heated filament. The gas introduced into the chamber is catalytically excited, or decomposed into radicals or ions, by a metal (Pt, W, Ta, etc.) filament heated to around 1800–2000 °C. Then, for example, Si radicals diffuse inside the chamber and are deposited onto a substrate that is maintained at a relatively high temperature (150–450 °C). Although device-quality films with high deposition rates are obtained using this technique, a drawback is the poor uniformity of deposited films.

Pulsed laser deposition

Pulsed laser deposition (PLD) is very simple when compared with sputtering and the various CVD techniques described so far. A pulsed high-power laser beam is focused onto a target in a vacuum chamber causing evaporation of the target source material. This resembles electron-beam evaporation, but in PLD a directional plasma plume is created by the absorption of photons, and hence this method is distinguished from simple thermal evaporation (Delahoy and Guo, 2010). The pulse energy is 300–600 mJ $pulse^{-1}$, and the instantaneous power density reaches 10^9 W cm^{-2}. Excimer gas (KrF or ArF) lasers are used for this purpose, as are solid state Nd:YAG lasers. A substrate faces the target, and the ablated source material condenses on it and grows the film. Stoichiometric transfer of atoms from multicomponent source materials to the substrate is performed by the PLD technique. It is now widely used to create device-quality transparent and conductive oxide (TCO) films, in particular commercially available IGZO films (Kamiya *et al.*, 2009).

2.2 Melt-quenched glasses

Most oxide and chalcogenide glasses can be prepared via the melt-quenched (MQ) method. Techniques used to prepare glass fibers are, in principle, based on the MQ method. When the temperature of the melt (the liquid state) decreases, melts become *supercooled liquids*, and a further decrease in temperature below the *glass-transition temperature* T_g produces glassy materials. The terms glass or *glassy material* are usually applied to materials prepared via the MQ method. Thus, glasses prepared by the MQ method undergo the glass transition. It is of interest to note that the empirical relation $T_g \approx 2T_m/3$ is reported in most glasses (Mott and Davis, 1979), where T_m is the *melting temperature*. When a glassy material is heated, crystallization occurs at a temperature T_c between T_g and T_m. In the laboratory, examples of chalcogenide glasses are sealed in quartz ampoules at 10^{-6} Torr and heated beyond their melting temperature. The samples are agitated (through rotation and vibration) then cooled quickly by immersion in cold water or air (at ambient temperature). Glasses are formed due to this high cooling rate (Elliott, 1990; Tanaka and Shimakawa, 2011). The ability under normal conditions of materials to form glasses depends also on the compositions of the materials (see, for example, Borisova (1981) and Popescu (2000)).

The nature of the glass transition is an important physical parameter for melt-quenched glasses (Elliott, 1990; Tanaka and Shimakawa, 2011), and hence T_g will be discussed in more detail in Chapter 3. It should be also noted that T_g and T_c are important parameters for phase-change materials, and therefore a detailed discussion, together with phase-change behavior, will also be given in Chapter 3.

2.3 Other techniques

There are other techniques that produce amorphous states (Elliott, 1990). One is irradiation-induced amorphization, which we introduce briefly here. The interaction between high-energy ionizing particles and crystalline solids can produce heavy structural damage, leading eventually to amorphous states. We know, for example, that almost all a-Si films made by the deposition techniques from gas states mentioned in Section 2.2 contain significant voids that may extend to several nanometers. These voids disturb the analysis of physical properties such as the radial distribution function (RDF), and hence pure and void-free a-Si is required.

To achieve this, we employ self-ion implantation at MeV energies of crystalline silicon (c-Si) at 77 K, i.e. Si itself is implanted into c-Si to produce pure and void-free a-Si (Laaziri *et al.*, 1999). After bombardment, the Si wafer is chemically etched from the back surface to remove c-Si; an amorphous layer of approximately 10 μm is obtained with this method. Similar implantations have been performed in group III–V compounds such as GaAs, GaSb, InAs, and InSb, with thicknesses around 2.5 μm obtained (Ridgway *et al.*, 2003). Amorphization of c-Ge substrates under ion implantation (up to $\sim 10^{16}$ cm^{-2}) has also been studied using transmission electron microscopy (TEM) (Koffel *et al.*, 2006) and Rutherford backscattering spectroscopy (RBS) (Impellizzeri,

Mirabella, and Grimaldi, 2011). It is known that the Ge surface undergoes a topology change and the cellular structure appears.

Finally, we introduce a special preparation method that is not popular in this field: mechanical milling (MM) (Koch *et al.*, 1983; Schwarz and Koch, 1986) in the solid state. The rapid quenching technique is based on the rapid removal of kinetic energy in an energized state. Mechanical milling comprises a homogenization process up to the atomic level by chemical reaction in a system with a negative heat of mixing. In fact, the amorphization of trigonal Se can proceed by mechanical milling (Fukunaga *et al.*, 1996). The number of atoms located at the second-nearest-neighbor distance ($r = 0.34$ nm) in the RDF is known to be reduced from four to zero, leading to the conclusion that the bonding between chain molecules in trigonal Se is destroyed by the strain and defects induced by the milling. Interestingly, a-Se prepared by MM exhibits a glass transition at the same temperature (320 K) as that occurring in MQ-Se, whereas crystallization occurs at a lower temperature (~370 K) than that for MQ-Se (~400 K).

Mechanical milling proceeds through the destruction of the crystal structure of an intermetallic compound or pure element due to an accumulation of strains and defects. Note that as the starting material (before MM) is energetically in the lowest (crystalline) state, the "uphill" process should be involved in the initial stage, during which energy is given mechanically to the material system. Chain molecules will be destroyed first by the strain, and defects are induced at this stage.

2.4 The Phillips constraint theory

The formation of glasses is highly dependent on network constraints, and hence we discuss the theory here. According to the Phillips constraint theory (Phillips, 1979), ideal glass-forming conditions occur when the average number of inter-atomic force-field constraints equals the number of degrees of freedom per atom. An atom that has all covalent bonds satisfied obeys Mott's $(8 - N)$ rule (Mott, 1969), i.e. Se has $N_c = 2$, As and P have $N_c = 3$, Si has $N_c = 4$, etc. For a binary alloy A_xB_{1-x}, the average coordination is given by

$$Z = xN_c(\mathrm{A}) + (1 - x)N_c(\mathrm{B}), \tag{2.1}$$

where $N_c(\mathrm{A})$ is the coordination number of atom A. The number of stretched bonds is equal to $Z/2$ because every sigma bond belongs to two atoms. A new bond direction is derived by the addition of two bond angles. Therefore the bond-bending number is $2Z - 3$. The total number of linearly independent constraints $N_c = Z/2 + (2Z - 3)$. By the Phillips theory, N_c must be equal to the degrees of freedom. In the three-dimensional (3D) case, $N_c = N_d = 3$, so

$$3 = Z/2 + (2Z - 3), \tag{2.2}$$

and we obtain $Z = 2.4$.

In systems such as a-GeS, a-GeSe, a-SiS, a-SiTe, etc., group IV elements are connected to four neighbors and group VI elements are connected to two. For a-$A^{IV}{}_x B^{VI}{}_{(1-x)}$, we have $4x + 2(1 - x) = 2.4$, so $x = 0.2$, i.e. a-GeS_4, a-$GeSe_4$, a-SiS_4, a-$SiSe_4$, and a-$SiTe_4$ are the optimum and mechanically most stable compositions. It is noted again that GeS_2 is the chemically most stable composition. In group V–VI binary alloy systems such as a-AsS, a-AsTe, etc. (group V elements have three neighbors), we obtain $x = 0.4$ from $3x + 2(1 - x) = 2.4$, and we conclude that a-As_2S_3, a-As_2Se_3, etc. are the optimum compositions. For amorphous semiconductors with $Z < 2.4$, the structure is "underconstrained" or "floppy" and for $Z > 2.4$ the structure is "overconstrained" or "rigid".

Some chalcogenide glasses have an average coordination number of $Z = 2.67$. The constraint for an atom in the two-dimensional (2D) plane ($N_d = 2$) is defined as $N_c = Z/2 + (Z - 1) = 3$ because a new sigma bond is defined by one angle (Tanaka, 1989). The following general equation may be written for 2D and 3D cases:

$$N_c = \frac{Z}{2} + \frac{(N_d - 1)(2Z - N_d)}{2}. \tag{2.3}$$

Then we obtain a general expression for the average coordination number as follows:

$$Z = \frac{2N_c + (N_d - 1)N_d}{1 + 2(N_d - 1)}. \tag{2.4}$$

The SiO system is a well-known exception to the Phillips theory, with SiO_2 instead of SiO_4 being the optimum glass composition. This is because the Si–O–Si bond-angle distribution is rather wide (Kuo, Lee, and Hwang, 2008), and therefore the constraint associated with oxygen bond angles is weak and may be neglected. Consider a $Si_xO_{(1-x)}$ binary alloy. In the equation $3 = Z/2 + (2Z - 3)$, the $(2Z - 3)$ term associated with the bond angles must be modified as there is no zero-bond-angle constraint for oxygen. The revised term for bond bending is therefore

$$(2Z_{Si} - 3)\,x + 0\,(1 - x) = 5x$$

(because $Z_{Si} = 4$). We must thus solve the following equation:

$$3 = Z/2 + 5x,$$

where the average coordination number is again given by $Z = 4x + 2(1 - x)$. We obtain $x = 1/3$, and hence SiO_2 is the best glass-forming composition.

References

Borisova, Z.U. (1981). *Glassy Semiconductors*. New York: Plenum.

Delahoy, A.E. and Guo, S. (2010). Transparent conducting oxides for photovoltaics. In *Handbook of Photovoltaic Science and Engineering*, 2nd edn., eds. A. Luque and S. Hegedus. Chichester: John Wiley and Sons, pp. 716–796.

Elliott, S.R. (1990). *Physics of Amorphous Materials*, 2nd edn. Harlow: Longman Science & Technical.

Fukunaga, T., Utsumi, M., Akatsuka, H., Misawa, M., and Mizutani, U. (1996). Structure of amorphous Se prepared by milling. *J. Non-Cryst. Solids*, **205**–207, 531–535.

Impellizzeri, G., Mirabella, S., and Grimaldi, M.G. (2011). Ion implantation damage and crystalline-amorphous transition in Ge. *Appl. Phys. A*, **103**, 323–328.

Kamiya, T., Nomura, K., and Hosono, H. (2009). Origins of high mobility and low operation voltage of amorphous oxide TFTs: electronic structure, electron transport, defects and doping. *J. Disp. Technol.*, **5**, 468–482.

Koch, C.C., Calvin, O.B., Macklamey, C.G., and Scarbrough, J.O. (1983). Preparation of ''amorphous'' $Ni_{60}Nb_{40}$ by mechanical alloying. *Appl. Phys. Lett.*, **43**, 1017–1019.

Koffel, S., Claverie, A., BenAssayag, G., and Scheibin, P. (2006). Amorphization kinetics of germanium under ion implantation. *Mater. Sci. Semicond. Process.*, **9**, 664–667.

Kuo, C-L., Lee, S., and Hwang, G.S. (2008). Strain-induced formation of surface defects in amorphous silica: a theoretical prediction. *Phys. Rev. Lett.*, **100**, 076104, 1–4.

Laaziri, K., Kycia, S., Roorda, S. *et al.* (1999). High resolution radial distribution function of pure amorphous silicon. *Phys. Rev. Lett.*, **82**, 3460–3463.

Mahan, A., Xu, Y., Nelson, B. *et al.* (2001). Saturated defect densities of hydrogenated amorphous silicon grown by hot-wire chemical vapour deposition at rates up to 150 Å. *Appl. Phys. Lett.*, **78**, 3788–3790.

Mott, N.F. (1969) Conduction in non-crystalline materials. *Phil. Mag.*, **19**, 835–852.

Mott, N.F. and Davis, E.A. (1979). *Electronic Processes in Non-Crystalline Materials*, 2nd edn. Oxford: Clarendon Press.

Phillips, J.C. (1979). Topology of covalent non-crystalline solids I: short–range order in chalcogenide alloys. *J. Non-Cryst. Solids*, **34**, 153–181.

Pócsik, I., Veres, M., Füle, M. *et al.* (2003). Carbon nano-particles prepared by ion-clustering in plasma. *Vacuum*, **71**, 171–176.

Popescu, M.A. (2000). *Non-Crystalline Chalcogenides*. Dordrecht: Kluwer.

Ridgway, M.C., Azevedo, G. de M., Glover, C.J., Yu, K.M., and Foran, G.J. (2003). Common structure in amorphised compound semiconductors. *Nucl. Inst. Meth. Phys. Res. B*, **199**, 235–239.

Rowlands, J.A. and Kasap, S.O. (1997). Amorphous semiconductors usher in digital X-ray imaging. *Phys. Today*, **50**, 24–31.

Schiff, E.A., Hegedus, S., and Deng, X. (2010). Amorphous silicon-based solar cells. In *Handbook of Photovoltaic Science and Engineering*, 2nd edn., eds. A. Luque and S. Hegedus. Chichester: John Wiley and Sons, pp. 487–545.

Schwarz, R.B. and Koch, C.C. (1986). Formation of amorphous alloys by the mechanical alloying of crystalline powders of pure metals and powders of intermetallics. *Appl. Phys. Lett.*, **49** 146–148.

Shimakawa, K., Hayashi, K., Kameyama, T., and Watanabe, T. (1991). Anomalous electrical conduction in graphite-vaporized films. *Phil. Mag. Lett.*, **64**, 375–378.

Spear, W.E. and LeComber, P.G. (1975). Substitutional doping of amorphous silicon. *Solid State Commun.*, **17**, 1193–1196.

Tanaka, K. (1989). Structure phase transitions in chalcogenide glasses. *Phys. Rev. B*, **39**, 1270–1279.

Tanaka, K. and Shimakawa, K. (2011). *Amorphous Chalcogenide Semiconductors and Related Materials*. New York: Springer.

Tanioka, K. (2007). The ultra sensitive TV pickup tube from conception to recent development. *J. Mater. Sci.*, **18**, S321–S325.

Veres, M., Füle, M., Tóth, S. *et al.* (2005). Simultaneous preparation of amorphous solid carbon films, and their cluster building blocks. *J. Non-Cryst. Solids*, **351**, 981–986.

Veres, M., Koós, M., and Pócsik, I. (2002). IR study of the formation process of hydrogenated amorphous carbon film. *Diam. Relat. Mater.*, **11**, 1110–1114.

Veres, M., Tóth, S., and Koós, M. (2008). New aspects of Raman scattering in carbon-based amorphous materials. *Diam. Relat. Mater.*, **17**, 1692–1696.

3
Structure

A brief summary of the essential differences between amorphous and crystalline semiconductors is presented at the beginning of this chapter for readers who are unfamiliar with the amorphous phase. The amorphous phase is not the lowest energy state, and these materials tend to relax to a crystalline phase, which is structurally the lowest energy state. Different materials need different time scales to reach the ground state. Amorphous structures at an atomic scale that have a large relaxation time are discussed in the main part of this chapter. At the end of the chapter, we discuss materials which are able to change their phases within a nanosecond time scale.

3.1 Differences between amorphous and crystalline semiconductors

Charge distributions

Before we enter into a detailed discussion of diffraction patterns, we present a brief overview of some other significant measurable differences between crystalline and amorphous phases, including both non-destructive and destructive methods. Atoms in monoatomic crystalline semiconductors have no net atomic charge at their equilibrium positions because of their symmetric arrangement, but they do carry charges in the amorphous case due to geometric distortions. These charge accumulations play an important role in the chemical shift of nuclear magnetic resonance (NMR) measurements for high-resolution core-level spectra, or for infrared absorptivity. In the 1980s, this phenomenon was one of the most intensively investigated topics in amorphous semiconductor physics (Guttman, Ching, and Rath, 1980; Klug and Whalley, 1982; Ley, Reichardt,

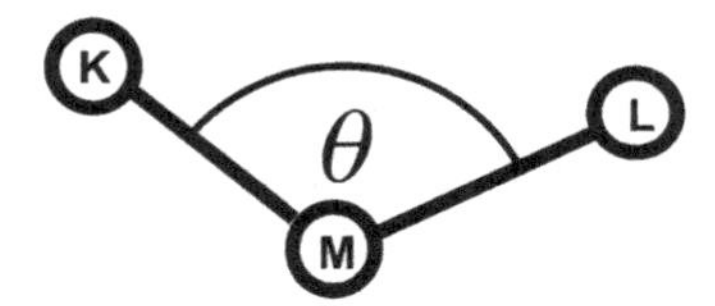

Figure 3.1. An elementary triad of atoms, denoted by K, L, and M, forming two bonds (KM and LM), with the bond angle KLM denoted by θ.

and Johnson, 1982; Kramer, King, and Mackinnon, 1983; Brey, Tejedor, and Verges, 1984; Winer and Cardona, 1986; Kugler and Náray-Szabó, 1987; Bose, Winer, and Andersen, 1988; Kugler, Surjan, and Náray-Szabó, 1988; Kugler and Náray-Szabó, 1991a, 1991b).

Net atomic charges are not observable directly, but experimental determination of their fluctuation is possible. By considering the deviation from the ideal bond angle, a simple model was developed for the derivation of net atomic charges in a-Si and diamond-like a-C (da-C) (Kugler *et al.*, 1988; Kugler and Náray-Szabó, 1991b). Consider an elementary triad of atoms, denoted by K, L, and M, forming two bonds (KM and LM), with the angle KML denoted by θ (see Figure 3.1). The net charge on the atoms of the triad depends linearly on the deviation of θ from the ideal tetrahedral value ($d\theta = \theta - 109.47°$):

$$q_{\mathrm{M}} = 2Ad\theta \quad \text{and} \quad q_{\mathrm{K}} = q_{\mathrm{L}} = -Ad\theta, \tag{3.1}$$

where A is a fitting parameter. The total atomic net charge on atom M, q_{M}^{total}, is a sum of the contributions originating from all triads containing atom M. Because in the distorted tetrahedral model each atom is at the center of 6 triads and at the end of 12 triads (see Figure 3.2), we obtain the following relation:

$$q_{\mathrm{M}}^{total} = A\left(2\sum_{i=1}^{6} d\theta_i - \sum_{j=1}^{12} d\theta_j\right). \tag{3.2}$$

From the semiempirical Hartree–Fock cluster calculation, $A = -0.69$ millielectrons/deg for a-Si (Kugler *et al.*, 1988) (Figure 3.3) and $A = -0.51$ millielectrons/deg for da-C (Kugler and Náray-Szabó, 1991b) have been obtained. In order to estimate charge fluctuations in a-Si and da-C, the geometric model of tetrahedrally bonded amorphous semiconductors proposed by Wooten, Winer, and Weaire (1985) was used. Applying eqn. (3.2) to all 216 atoms of the

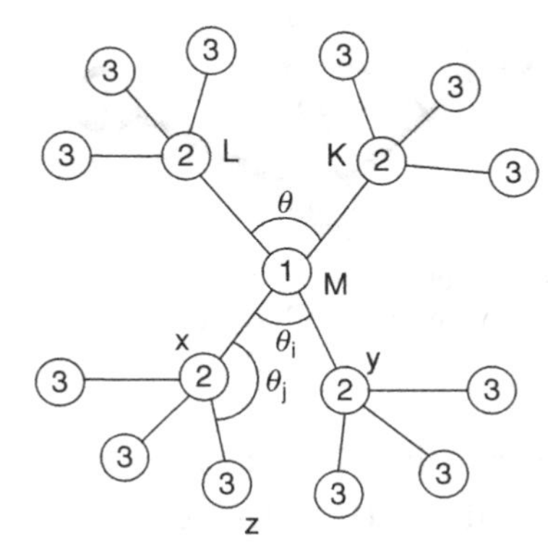

Figure 3.2. Topological model for net atomic charge calculation.

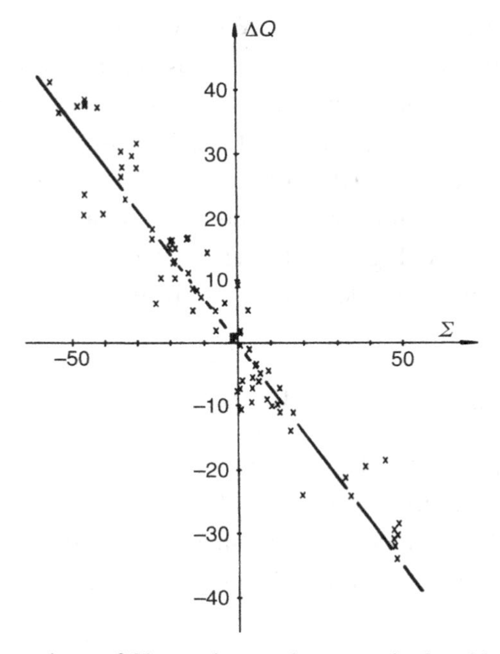

Figure 3.3. Comparison of Si atomic net charges calculated by semiempirical Hartree–Fock calculation (crosses) and by eqn. (3.2) (solid line). Here Σ denotes the quantity in parentheses in eqn. (3.2). Net charge is in millielectrons, Σ is in degrees. (Taken with permission from Kugler *et al.* (1988). *Phys. Rev. B*, **37,** 9069. Copyright 2013 by the American Physical Society.)

model cluster, $dq_{\mathrm{Si}} = 0.021$ electrons for the root mean square (rms) charge fluctuation in a-Si and $dq_{\mathrm{C}} = 0.015$ electrons in da-C have been obtained. A revised analysis of a core-level spectroscopic measurement (Ley *et al.*, 1982) yielded the estimation that the charge fluctuation in a-Si must be lower than 0.030 electrons. Recently, a similar theoretical model was reported for distorted long disordered

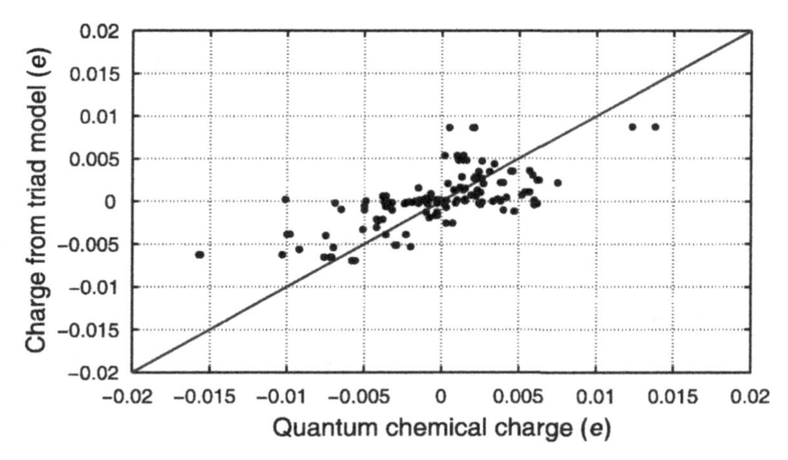

Figure 3.4. Comparison of Se atomic net charges calculated using the Hartree–Fock *ab initio* calculation and eqn. (3.4). Each symbol represents an atom. (From Lukács and Kugler (2010). *Chem. Phys. Lett.*, **494**, 287. Copyright 2013 with permission from Elsevier.)

selenium chains (Lukács and Kugler, 2010). The net charge on the atoms of a triad depends on the deviation of θ from the average value ($d\theta = \theta - 101°$):

$$q_M = 2\left(Ad\theta + Bd\theta^2\right) \quad \text{and} \quad q_K = q_L = -\left(Ad\theta + Bd\theta^2\right). \qquad (3.3)$$

The total atomic net charge on atom M, q_M^{total}, is a sum of the contributions originating from the three triads containing atom M:

$$q_M^{total} = 2\left(Ad\theta_M + Bd\theta_M^2\right) - A\left(d\theta_{M-1} + d\theta_{M+1}\right) - B\left(d\theta_{M-1}^2 + d\theta_{M+1}^2\right). \qquad (3.4)$$

Based on the Hartree-Fock *ab initio* calculation, $A = -0.45$ millielectron/deg and $B = -0.0089$ millielectron/deg^2 have been obtained as the best fits (Figure 3.4).

Heat capacity

In the crystalline case, the heat capacity is easy to derive from the lattice vibration model. A high-temperature limit applying both classical and quantum-mechanical models provides us with a value of $C = 3Nk_B$, where N and k_B are the number of atoms and the Boltzmann constant, respectively. Low-temperature heat capacity using a quantum-mechanical calculation is in good agreement with experiments, whereas the classical counterpart ($C = 3Nk_B$, the Dulong–Petit law) is not applicable. The total energy due to the lattice vibration is obtained by integrating the energy of a single quantum oscillator as follows:

$$E_{osc} = \int_0^\infty \left(\hbar\omega/2 + \frac{\hbar\omega}{e^{\hbar\omega/k_B T} - 1}\right) g(\omega)\, d\omega, \tag{3.5}$$

where ω is the angular frequency ($\hbar\omega$ is the phonon energy) and $g(\omega)$ represents the vibration density of states per unit frequency range, i.e. the number of modes with frequency $[\omega; \omega + d\omega]$ is $g(\omega)d\omega$. Only lattice modes of low frequency will be excited at low temperatures. The dispersion relation has three acoustic branches, one longitudinal and two transverse:

$$g(\omega) = \frac{V\omega^2}{2\pi^2}\left(\frac{1}{v_L^3} + \frac{2}{v_T^3}\right), \tag{3.6}$$

where v_L and v_T are the sound velocities of two different modes. The heat capacity due to phonon excitation at low temperature is given by

$$C = \frac{dE_{osc}}{dT} = \frac{2V\pi^2 k_B}{15}\left(\frac{1}{v_L} + \frac{2}{v_T}\right)\left(\frac{k_B T}{\hbar}\right) \sim T^3. \tag{3.7}$$

In addition, the contribution due to the electrons must be considered for a more accurate calculation of heat capacity. The most general model is the free Fermi-electron-gas description. The low-temperature approach obtained from a Bethe–Sommerfeld expansion yields

$$C_{el} = \frac{\pi^2}{2} N k_B \frac{T}{T_F} \sim T, \tag{3.8}$$

where T_F is the Fermi temperature. Therefore the total heat capacity of crystalline materials can be written as

$$C_{total} = \gamma T + \beta T^3. \tag{3.9}$$

This model does not work for amorphous semiconductors; there is no low-frequency collective excitation in amorphous materials. Another problem is that the free electron model is applicable for metals, but it does not work well for semiconductors. Near linear heat capacity has been observed at low temperatures for a large class of different materials, including chalcogenide glasses. The common property among these structures is the low average coordination number of atoms. In the same year, Phillips (1972) and Anderson, Halperin, and Varma (1972) proposed the "two-level system" model for the description of linear dependence of low-temperature heat capacity. The starting point is a fused asymmetric double-well potential having two different bottom points. Atoms are able to occupy one of the two equilibrium positions and to tunnel to the other position. The free energy of this classical statistical basic problem having two energy states in a canonical ensemble is given by

$$F(T, V, N) = U - TS = -k_B T \ln(\cosh(\Delta E/2k_B T)), \tag{3.10}$$

where ΔE is the energy difference between the two states. The heat capacity can be calculated using

$$C_V = -T\left(\frac{\partial^2 F}{\partial T^2}\right)_V. \tag{3.11}$$

For a two-level system,

$$C_V = (\Delta E/2k_B T)^2 \mathrm{sech}^2(\Delta E/2k_B T). \tag{3.12}$$

At low temperatures (around 1 K), $k_B T \approx 10^{-4}$ eV, which can be considered as an upper limit for excitation energies, E_{max}. As a first approach we use a constant density of states, i.e. $n(\Delta E) = n$. The total heat capacity can be obtained as follows:

$$C_{total} = \int_0^{E_{max}} n\left(\frac{\Delta E}{2k_B T}\right)^2 \mathrm{sech}^2\left(\frac{\Delta E}{2k_B T}\right) d\Delta E. \tag{3.13}$$

As sech is a rapidly decreasing function ($\sim e^{-x}$), we put infinity instead of E_{max} and the integral calculus can be carried out analytically:

$$C_{total} = \frac{\pi^2}{6} k_B^2 n T \sim T. \tag{3.14}$$

We can conclude that the total heat capacity of amorphous semiconductors at low temperatures is proportional to T.

Calorimetric properties (glass transition, crystallization)

A calorimetric measurement, such as differential scanning calorimetry (DSC), monitors a change in the specific heat of a material. When a glass sample is heated from low temperatures, heat is absorbed at the glass-transition temperature T_g (an endothermic process) and is emitted at the crystallization temperature T_c (an exothermic process). *Amorphous* materials do not show such an endotherm below T_c in the DSC measurement. An understanding of the mechanism of crystallization is very important for phase-change materials, and this topic will be discussed in Section 3.4 in detail. Above T_c an endothermic reaction again occurs at the melting temperature T_m. As the glass transition and crystallization are fundamentally important subjects in glass science, details of these phenomena are discussed in the following sections. Note that crystallization is directly related to phase-change devices, and therefore this will also be discussed in Section 3.4.

Glass-transition temperature

As discussed in Chapter 2, glasses prepared by the melt-quenched (MQ) method undergo a glass transition at T_g, which is defined as the transforming temperature between the glassy and supercooled liquid states. A supercooled liquid describes a material that remains in a liquid state, even below T_m. Although the definition of the glass transition itself has a clear meaning, the experimental determination of T_g is not easy, because the deduced T_g depends on experimental conditions, such as the cooling rate of the melt in preparation and/or the heating rate in the DSC measurement itself. It is still unknown which factors dominate T_g (Elliott, 1990; Greaves and Sen, 2007; Tanaka and Shimakawa, 2011). In spite of this ambiguity in the determination of T_g, it does correlate to some physical parameters. A well-known empirical rule is $T_g \approx 2T_m/3$ (where the temperatures are measured in kelvin) (Kauzmann, 1948). Recently, it has been suggested that the quantum effect plays a role in the glass-transition temperature (Novikov and Sokolov, 2013): quantum effects lead to a significant decrease in T_g with respect to T_m, so that the ratio T_g/T_m can be much smaller than 2/3 in materials where T_g is near or below 60 K (lower temperatures), and is given by

$$\frac{T_g}{T_m} = \frac{A}{(1 + B/T_g)}, \tag{3.15}$$

where $A \approx 2/3$ and B depends on the strength of the Boson peak and some other parameters.

As T_m can be primarily related to a cohesive energy, T_g should also be related to the cohesive (or bond) energy in covalent glasses; this is discussed in the following. At the glass-transition temperature, a glassy matrix may be destroyed; T_g is therefore expected to be related to a network constraint (or coordination number). Here, we introduce some representative models on this topic, although there are numerous approaches (see, for example, Tichý and Tichá, 1995). Tanaka (1985) has proposed the following empirical relation for

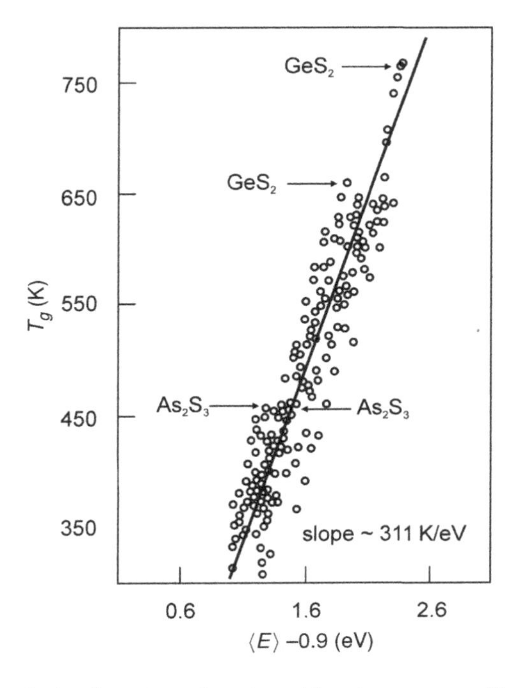

Figure 3.5. Correlation between glass-transition temperature T_g and the overall mean bond energy $\langle E \rangle$ of a covalent network of chalcogenide glasses. (From Tichý and Tichá (1995). *J. Non-Cryst. Solids*, **189**, 141. Copyright 2013 with permission from Elsevier.)

some covalent bonding glasses involving polymers: $\ln T_g \approx 1.6Z + 2.3$, where Z is the average coordination number of the atoms. The parameter Z is known to be important (see Section 2.4) in the description of the physical and chemical properties of glasses, and there exist the "magic" numbers $Z = 2.4$ and 2.67 in the constraint theory of glassy networks (Tanaka and Shimakawa, 2011). For the vast majority of chalcogenide glasses, however, T_g is not simply related to Z (Tichý and Tichá, 1995; Freitas, Shimakawa, and Kugler, 2013).

Instead of simply using Z, by taking the overall mean bond energy, $\langle E \rangle = \Sigma_I f_i E_i Z_i$, the other empirical relation $T_g = 311(\langle E \rangle - 0.9)$ K, as shown in Figure 3.5, has been proposed for covalent-bonding chalcogenide glasses (Tichý and Tichá 1995). Here, f_i is the atomic fraction of the ith component and Z_i is the coordination number of the ith atom. Note that some special chalcogenides, such as molecular-type glassy materials, are excluded from the data points because $\langle E \rangle$ does not involve intermolecular interaction. The fact that T_g is well scaled with $\langle E \rangle$ in covalent chalcogenide glasses suggests that T_g is principally determined

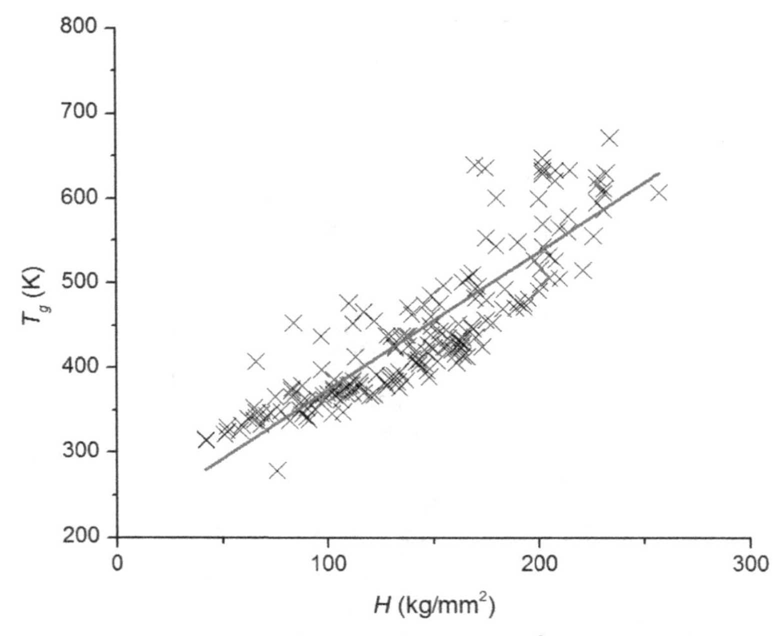

Figure 3.6. Correlation between T_g(K) and H(kg/mm^2) in chalcogenide glasses. Solid line is least-square fit to the data, producing $T_g = 1.6H + 211$. (From Freitas, Shimakawa, and Kugler (2013). *Chalcogenide Lett.*, **10**, 39. Copyright 2013.)

by the cohesive energy. The short-range structure of glasses may be an important property for T_g.

The average bond energy is not a measurable quantity and hence, instead of scaling with $\langle E \rangle$, the other measurable parameters should be used for scaling. The hardness H may correlate with $\langle E \rangle$, which is never applied as a scaling parameter to express T_g. Principally, H can be related to T_m in a covalent-bond system, and T_g vs. H may have a linear relationship. Figure 3.6 shows a good correlation between T_g and H in a covalent amorphous chalcogenide system. The scatter of the experimental data suggests that the intermolecular (van der Waals) interaction should not be ignored (Freitas, Shimakawa, and Kugler, 2013).

Finally, we mention that computer simulations are also useful in understanding the nature of the glass transition (Langer, 2006; Wilson and Salmon, 2009).

Viscosity

Glasses experience a supercooled liquid state when they are cooled below their melting points. Glass-forming liquids have high viscosities even at high temperatures, and the viscosity of these liquids changes rapidly near the glass-transition temperature. Thus the temperature at which the viscosity reaches a certain value (10^{12} Pa s) is also used as a determination of T_g. The empirical rule for temperature-dependent viscosity η, called the Vogel–Tammann–Fulcher equation, in various glasses is given as $\eta = \eta_0 \exp[A/(T - T_0)]$, where A and T_0 are constants (Angell, 1988, 1995); T_0 is related to the *fragility* of glasses. Strong glasses, for example SiO_2, have $T_0 = 0$ (following an Arrhenius relation), and the fragile structure shows non-Arrhenius behavior. Note that the viscosity is continuous through T_g, whereas the heat capacity shows a discontinuity (remember the DSC measurement). The above form of the relation appears in many transport phenomena, such as ionic conduction in liquid states. We will see this empirical law in Section 3.4 on phase-change phenomena.

The model proposed by Adam and Gibbs (1965) is widely accepted in explaining temperature-dependent viscosity, in which the important factor determining the relaxation phenomena is the configurational entropy (CE). Alternatively, Aniya (2002) has suggested that fluctuations in the coordination number ΔZ and the binding energy ΔE in a glassy system dominate the fragility of glass: large fluctuations in ΔZ and ΔE produce an increased fragility.

3.2 Projection from three-dimensional structures to one-dimensional functions

Information about the atomic-scale structure of materials is essential in the derivation and understanding of different physical properties. The distribution of atoms as a function of distance, i.e. the probability of finding a particle at distance r when another particle can be located at the origin, is a possible characterization of the geometric structure at the microscopic level. In the crystalline case, diffraction experiments produce diffractograms with the characteristic Bragg reflections evidenced by well-defined sharp peaks at given angular positions. As a typical example, peak positions and their coordination numbers Z within a 0–1.0 nm interval for crystalline silicon are displayed in Table 3.1. Pure amorphous semiconductors do not possess the long-range translational order of atomic sites and therefore have no discrete diffraction patterns.

There are two one-dimensional functions that are used to describe three-dimensional atomic distributions. In the first function, the atomic distribution in a well-defined direction is investigated. In this case, the pair correlation (distribution) function $g(r)$ is a convenient function, where $g(r)dV$ is proportional

Table 3.1. *Peak positions and their coordination numbers within a 0–1.0 nm interval for crystalline silicon.*

	Peak position (nm)	Z
1	0.235	4
2	0.384	12
3	0.45	12
4	0.543	6
5	0.591	12
6	0.665	24
7	0.705	16
8	0.768	12
9	0.803	24
10	0.858	24
11	0.89	12
12	0.94	8
13	0.969	24

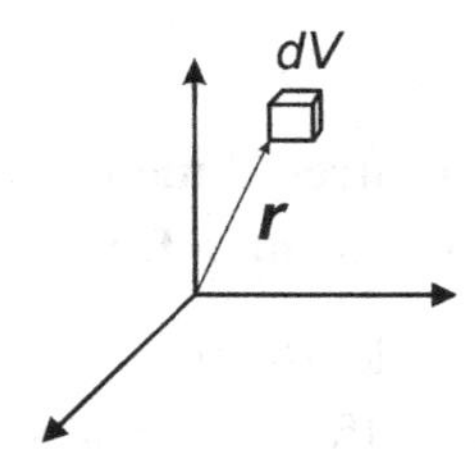

Figure 3.7. Schematic for the pair correlation function. Vector ***r*** indicates the direction along which the atomic distribution is calculated.

to the probability of finding a particle inside a volume dV (see Figure 3.7). The radial distribution function (RDF) is also a measure of the probability of finding an atom a distance r away from a reference particle. This function, $J(r)$, represents the number of atoms within a distance between r and $r + dr$ from a given particle (see Figure 3.8). In contrast to structures that have translation symmetry, amorphous materials are isotropic; therefore the pair correlation function and the RDF contain the same information because of the following relationship: $J(r)dV = g(r)4\pi r^2\, dr$. Functions must be zero below a given distance, which is related to the atomic repulsion, and they should exhibit alternating maxima and minima following the coordination shells. The correlation with the atom

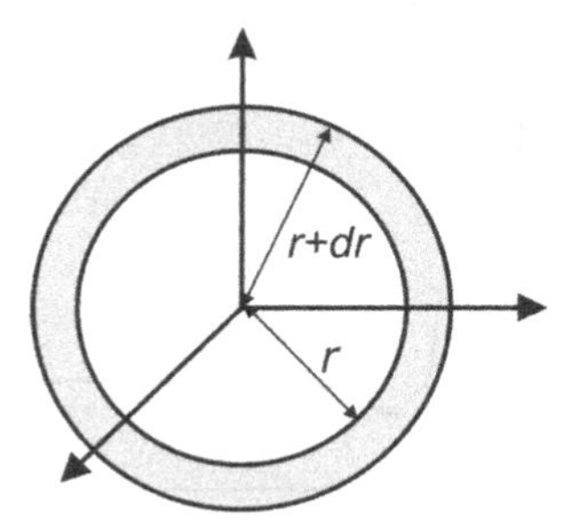

Figure 3.8. Schematic for the radial distribution function (RDF).

at the origin is lost at large r. It follows that $g(r)$ tends to unity (or to the density, depending on the normalization), whereas the RDF has a parabolic shape when r goes to infinity. Scientists working with amorphous materials usually use $J(r)$. The peak positions in $J(r)$ provide the radii of the different coordination shells of atoms surrounding the average atom. The area under the peaks derives the coordination numbers of the different shells of atoms. In a structural model, the RDF may be easily determined by calculating the distance between all the atomic pairs and putting them into a histogram. An alternative form of the RDF that is sometimes used is $T(r)$, which is equal to $J(r)/r$.

The structures of amorphous semiconductors are typically determined by neutron, electron, or x-ray diffraction experiments. Neutrons, having no charge, interact with the nucleus of each atom, whereas x-rays or electrons are sensitive to the electron distribution inside the amorphous materials. A neutron beam is capable of penetrating well beyond the surface of a sample, to depths on the order of centimeters in the condensed phases. In a diffraction experiment, we measure the intensity as a function of the scattering angle. Thermal neutrons with wavelengths of 0.1–0.2 nm are very useful for atomic-scale structural (and magnetic structure) investigations. Neutron diffraction is a non-destructive technique; it is a form of elastic scattering in which the neutrons exiting the sample have the same energy as the incident neutrons. In a standard neutron-diffraction measurement, $g(r)$ and $J(r)$ are derived from the measured scattered intensity by Fourier transformation of the static structure factor $S(\boldsymbol{Q})$, where $\boldsymbol{Q} = \boldsymbol{k}_{initial} - \boldsymbol{k}_{final}$. Here, $\boldsymbol{k}_{initial}$ and $\boldsymbol{k}_{final}$ represent the wave vectors in the initial and final states, and

$$\rho\left(g\left(\mathbf{r}\right)-1\right)=\frac{1}{(2\pi)^{3}}\int d\boldsymbol{Q}\left(S\left(\boldsymbol{Q}\right)-1\right)e^{-i\boldsymbol{Q}r}. \qquad (3.16)$$

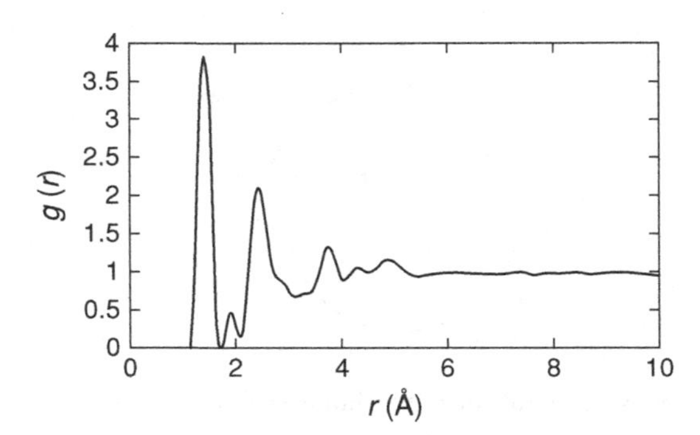

Figure 3.9. Pair correlation function of graphite-like a-C measured by neutron diffraction and derived from $S(Q)$ by Fourier transformation.

It is assumed in amorphous materials that the scattering is isotropic, and that this three-dimensional Fourier transformation can be reduced to the following one-dimensional sine transformation:

$$\rho\,(g\,(r) - 1) = \frac{1}{2\pi^2 r} \int_0^\infty Q\,[S\,(Q) - 1]\sin(Qr)\,dQ, \tag{3.17}$$

where $Q = (4\pi/\lambda)\sin\theta$ and r are already scalar values, λ is the de Broglie wavelength of the neutron, and 2θ is the scattering angle. A special example for a-C films is shown in Figure 3.9.

Neutron cross-sections can be divided into two parts: coherent and incoherent cross-sections, σ_{coh} and σ_{inc}, respectively. The elastic part of the coherent scattering provides information about the structure. Measured cross-sections for any element of the periodic table can be found in Sears (1992), pp. 29–37, or at http://www.ncnr.nist.gov/resources/n-lengths. The coherent, incoherent, and absorption cross-sections of isotopes are given in units of barn (1 barn = 10^{-28} m^2). In monatomic structure investigation, the absorption cross-section has no role. Materials with a larger coherent cross-section than their incoherent counterparts provide a good opportunity to derive the RDF experimentally using neutron diffraction.

X-rays are scattered by the electrons in atoms. When a covalent bond is formed, the associated valence electrons are distributed between the two atoms. It follows that the x-ray diffraction pattern probes the electron distribution around each atom and between the atoms. In amorphous cases, electrons are placed asymmetrically, so x-ray scattering may provide a more complex and different density distribution than that produced by neutron diffraction.

In contrast to x-ray and electron diffraction, neutrons are scattered by nuclei, and there is no relationship with atomic number, i.e. different isotopes of the same element can have widely different cross-sections. This fact is a great advantage for two- and three-component (multicomponent) materials, in which similar samples having different isotopes are prepared and measured (known as the isotope substitution method). In essence, the idea is to measure the diffraction pattern from several identical samples which contain different amounts of isotopes for one or more of the components. Each diffraction pattern will provide different partial structure factors, depending on the precise values of the neutron scattering lengths for each isotope.

Other ways of obtaining information on partial structure factors include the combined use of x-ray and neutron diffraction, or the joint use of neutron, x-ray, and electron diffraction data. An excellent review has been published by Gabriel J. Cuello on the elementary introduction to the method of structure factor determination by neutron diffraction in non-crystalline materials (Cuello, 2008).

Diffraction measurements on amorphous carbon

A pioneering electron diffraction measurement on evaporated carbon film was reported more than half a century ago by a Japanese group (Kakinoki *et al.*, 1960). They concluded that the film is built up by randomly distributed diamond-like and graphite-like crystalline regions. These regions have no mutual orientation over the range of several angstroms.

Neutron-diffraction measurements have been performed on amorphous carbon at low temperatures (Kugler *et al.*, 1993b) in order to investigate the origin of the anomalous electrical conduction (Shimakawa *et al.*, 1991). Using an arc vaporization method, a 1.0 g pure a-C sample was prepared. The experiment was carried out at the 7C2 spectrometer installed on the hot source of the Orphée reactor at CEA Saclay, in France. A momentum transfer range of 5–160 nm^{-1} was covered. In the measured pair correlation function, the positions of the first four main peaks were 0.142 nm, 0.246 nm, 0.283 nm (third-neighbor cross-ring distance), and 0.373 nm. In a perfect graphite crystal, the first four nearest-neighbor distances are 0.142 nm, 0.246 nm, 0.284 nm, and 0.364 nm. It seems likely therefore that the sample contained a high proportion of threefold-coordinated carbon atoms, i.e. it is typically a graphite-like material, as shown in Figure 3.9. Beyond 0.6 nm, there is an absence of atomic structure; $g(r)$ has no appreciable medium-range order. The derived $S(Q)$ values exhibit small differences due to the variation of the temperature, but there was no significant difference between the radial distribution functions.

The structure of another a-C sample (20–30 mg, a very limited specimen mass for neutron diffraction) prepared by plasma-arc deposition has been determined by neutron diffraction at the twin-axis diffractometer, D4, at the Institut Laue-Langevin (ILL), Grenoble, France (Gaskell *et al.*, 1991). The conclusion is in opposition to the above-mentioned results: the structure factor and reduced radial distribution functions are similar to those for a-Si and a-Ge, indicating a high proportion of tetrahedral bonding. The first three main interatomic distances are very close to bond lengths in diamond crystal (0.152 nm and 0.153 nm; 0.250 nm and 0.252 nm; 0.296 nm and 0.296 nm). A fit to the data yields about 86% tetrahedral bonding.

Thick a-C films were prepared in a high-vacuum system of base pressure around 10^{-7} Torr, by rf sputtering on liquid-nitrogen-cooled Cu substrates. Raman measurements were performed to confirm the amorphous nature of the deposited films. A 0.8 g sample was used for neutron-diffraction measurements performed at 300 K in the Special Environment Powder Diffractometer of the Argonne National Laboratory's Intense Pulsed Neutron Source, IL, USA (Li and Lannin, 1990). The films were removed from the substrate by using dilute hydrochloric acid. The RDF was compared to theoretical models, and was found to exhibit qualitative agreement with a number of basic models, indicating predominant threefold bonding. The absence of a specific peak at the graphite intrahexagon distance (third-neighbor peak) indicates that structural models with such intermediate-range correlations are not correct for a-C. This distinguishes a-C from local two dimensionally ordered graphite-like materials.

Diffraction measurements on amorphous silicon

The first electron diffraction measurements were carried out by Moss and Graczyk (1969), and by Barna *et al.* (1977). At the first important stage of amorphous silicon structure computer modeling, the lack of a neutron-diffraction measurement caused a measured RDF of a-Ge to be applied to make a comparison between theoretical and experimental data (Wooten *et al.*, 1985).

Some years later, a nearly pure, evaporated a-Si sample (0.5 g) was prepared in the Central Research Institute for Physics, Budapest, Hungary, and a neutron-diffraction experiment was performed on this sample at the 7C2 spectrometer installed on the hot source of the Orphée reactor at CEA Saclay, in France

(Kugler *et al.*, 1989). Using an incident wavelength of $\lambda = 0.0706$ nm, the momentum transfer range of 5–160 nm^{-1} was covered. A second neutron-diffraction measurement on the same a-Si sample was performed using the D4 twin-axis diffractometer at the high flux reactor at the ILL, Grenoble (Kugler *et al.*, 1993a). The incident neutron wavelength of 0.04977 nm and the angular ranges of 1.5°–65° and 46°–131° covered by the two multi-detectors provide us with a larger momentum transfer range of 3.3–230 nm^{-1}. The results of the experiments show that the covalently bonded a-Si films are not completely disordered. The bonds between the atoms and the coordination numbers are similar to those for the crystalline phase. Compared with a perfect crystal, a-Si shows that the first- and second-neighbor peaks are broadened but the positions are the same, whereas the third peak disappears in the measured RDF. This fact is one of the most important properties of group IV amorphous semiconductors. The measured mean square width of the first peak suggests that the bond-length fluctuations are around 1–2%. The broadening of the second peak reflects around 10% bond-angle fluctuation. The absence of the third peak confirms that there is no characteristic dihedral angle.

Neutron-diffraction measurements were performed on an a-Si sample as-deposited and annealed at 600 °C at the Special Environment Powder Diffractometer of the Argonne National Laboratory Intense Pulsed Neutron Source, IL, USA (Fortner and Lannin, 1989). It was prepared by rf sputtering at low Ar pressure. The radial distribution functions indicate values of 108.4° and 108.6° for a tetrahedral angle for as-deposited and annealed films, respectively, which are both smaller than the ideal 109.47° angle. The estimated bond-angle widths for the samples are 9.7° and 11.3° for as-deposited and annealed films, respectively. The coordination number in the annealed sample was 3.90, and a value of 3.55 was estimated for the as-deposited sample, which is smaller than the expected value of 4.

In addition to neutron-diffraction measurements, the structure factor $S(Q)$ of high-purity a-Si membranes prepared by self-ion implantation was measured over a large extended Q range of 0.3–550 nm^{-1} by x-ray diffraction, using a Huber six-circle diffractometer at the Cornell High Energy Synchrotron Source, A2-wiggler beam line, Ithaca, USA (Laaziri *et al.*, 1999). The sample was kept at 10 K to minimize the thermal effects on the diffraction pattern. The mean values of the coordination numbers were observed as 3.79 for the as-implanted sample and 3.88 for the annealed sample. In this nanovoid-free sample, the density deficit of a-Si relative to crystalline Si (c-Si) is due to a fundamental under-coordination, i.e. dangling bonds may decrease the local density.

A combined investigation using the complementary steady-state and pulsed neutron sources was carried out on hydrogenated amorphous silicon (a-Si:H) (Bellissent *et al.*, 1989). The aim of this study was to provide accurate data concerning the short-range order in a-Si:H and a-Si:D (where D means deuterium) samples. The samples were prepared by sputtering and the glow-discharge method. Samples of mass 1.0–1.5 g were measured on the D4 twin-axis spectrometer at steady state in the ILL, Grenoble, and on the LAD spectrometer at the ISIS spallation neutron source, Rutherford Appleton Laboratory, Chilton, UK. The Si–H distances were identified as 0.148 nm and 0.321 nm, and those for Si–Si as 0.234 nm and 0.375 nm for first- and second-neighbor distances, respectively.

Diffraction measurements on amorphous germanium

A large volume, corresponding to 9.6 g, of evaporated amorphous germanium, was contained in an 11.7 mm diameter thin-walled vanadium can for the neutron-diffraction experiment. The experiment was carried out at the ILL, Grenoble, on the D4 twin-axis diffractometer (Etherington *et al.*, 1982). The neutron wavelength was 0.05057 nm. The observed nearest-neighbor coordinate number, $Z_1 = 3.68$, is significantly less than the value of 4 expected for tetrahedral bonding. The second peak yielded a coordination number $Z_2 = 12.11$, which is somewhat greater than that expected for tetrahedral bonding. The structure of a-Ge is accurately described in terms of a tetrahedrally bonded random network.

Diffraction measurements on amorphous selenium

Neutron-diffraction measurements have been performed using the SLAD instrument at NFL, Studsvik, Nyköping, Sweden (Jóvári, Delaplane, and Pusztai, 2003). Crystalline selenium powder (3 g) was ball milled under an argon atmosphere for 6 hr in a Spex mixer/mill. The milling procedure consisted of milling for 15 min followed by 45 min of rest to avoid heating, and then the cycle was repeated. The amorphous selenium powder was contained in a thin-walled vanadium container. The obtained experimental structure factor $S(Q)$ was compared with three earlier measured data within the interval 8–120 nm^{-1}. Furthermore, pair correlation functions were derived by unconstrained reverse Monte Carlo computer simulations. The result is presented in Section 3.3.

3.3 Three-dimensional structure derived from one-dimensional function

Knowledge of the three-dimensional atomic arrangement is an essential prerequisite to understand the physical and chemical properties of different materials. In the crystalline case, the determination of the structure means solving the atomic arrangement within the unit cell containing a few atoms. The structure of the crystal as a whole is then constructed by repeating the unit cell periodically. In the non-crystalline case, the lack of periodicity means that the structure cannot be determined in the same way as for crystalline materials.

Amorphous semiconductor structures have been studied over a long period of time using different diffraction techniques. Although several experimental structural studies have been made using x-ray and electron diffraction, the neutron-diffraction technique yields the most accessible information on these disordered structures. In recent years, NMR (nuclear magnetic resonance) spectroscopy and EXAFS (extended x-ray absorption fine structure) spectroscopy have become important element-specific tools for the derivation of local atomic arrangements. The structure factor $S(Q)$ measured in a limited Q range can be Fourier transformed, which yields the radial distribution function. The obtained result is only a one-dimensional representation of the spatial atomic distribution in three dimensions. This projection causes information loss because an unlimited number of different possible disordered atomic structures can display the same RDF. Unfortunately, no experimental technique has been discovered so far that can determine the microscopic atomic distributions inside the bulk. A pioneering step was taken by Tegze and Faigel (1996) by using x-ray holography. They demonstrated the efficacy of atomic-scale x-ray holography by obtaining direct images of the three-dimensional arrangement of strontium atoms in cubic perovskite, $SrTiO_3$. To obtain holographic images on an atomic scale, thermal neutrons were applied by Cser *et al.* (2002). Disordered atomic arrangements still remain a problem. Efforts to develop modeling techniques with which to analyze atomic resolution are therefore continually being made.

Atomic interactions

Classical empirical potentials

In order to prepare three-dimensional atomic structures of amorphous semiconductors, a transferable atomic interaction is required. The simplest versions of atomic interactions are the classical empirical potentials, which are computationally the cheapest method for structural modeling. The parameters describing the atomic interaction can be derived either from experimental observations or from advanced quantum-mechanical calculations. For semiconductor structures, simple Morse or Lennard–Jones potentials are not suitable because the directional nature of atomic bonding also needs to be taken into account. Therefore, the potential needs to be expressed by at least two- and three-body terms. When a simulation is performed, a truncation of the interaction, known as the cut-off distance, is usually applied. The term "short range" means that the total potential energy of a given particle is dominated by the interaction with neighboring atoms that are closer than the cut-off distance.

Keating potential

The Keating empirical potential (Keating, 1966) is a simple example for a three-atom local interaction which has macroscopically measurable parameters. It was originally designed to describe the defect-free diamond type of crystal, such as silicon or germanium, or diamond-structure carbon, in which the atoms have four nearest neighbors. The bond-stretching and bond-bending force constants, α and β, respectively, are determined from the fitting to the elastic properties of the crystal phase. The number of bonds is equal to $4N/2 = 2N$, and the number of atomic triads is equal to $(4N \times 3)/2 = 6N$, where N atoms form the geometric structure. For these materials, the bond length in an equilibrium state is equal to r_0 (0.155 nm, 0.235 nm, and 0.245 nm are for C, Si, and Ge, respectively), and the bond angle is the tetrahedral angle ($\Theta = \arccos(-1/3) = 109.47°$). In the Keating potential, bonding geometries which deviate from these equilibrium values are assigned an energy penalty:

$$U = \frac{3}{8r_0^2}\left[\frac{\alpha}{2}\sum_{ij}^{2N}\left(r_{ij}^2 - r_0^2\right)^2 + \beta\sum_{jik}^{6N}\left(r_{ij}r_{ik} + r_0^2/3\right)^2\right], \tag{3.18}$$

where r_{ij} is the vector pointing from atom i to atom j, and $r_{ij}\ r_{ik}$ is the scalar product of two vectors. This formula is practically a Taylor expansion where the linear terms vanish and only the quadratic terms are kept. The constants α and β are obtained from a fit to the elastic properties of the crystal. For silicon, values of $\alpha = 296.5$ eV nm^{-2} and $\beta = 0.285\alpha$ were obtained. The validity of a large deviation from the equilibrium values of bond length and bond angle is questionable. It is suitable only for small displacements from the ideal positions. In particular, it should not be used for very disordered systems. Another disadvantage is that the adjacency matrix of a topological structure must be known for the total energy calculation. If the adjacency matrix is unknown, we cannot calculate the diffusion or other dynamical processes.

Several modified versions of the Keating potential, eqn. (3.18), have been developed. Fullerenes, nanotubes, and graphenes are formed by threefold-coordinated carbon atoms. Overney, Zhong, and Tománek (1993) reparameterized the potential, i.e. the bond-stretching and bond-bending force constants (α, β) were determined by applying the density-functional-theory (DFT) *ab initio* local-density approximation total energy calculations for specific distortion modes of graphite. The bond angle is equal to $\Theta = 120°$, and instead of the last term, $r_0^2/3$, we must substitute the term $r_0^2/2$. The structural rigidity and the low-frequency vibration modes of carbon have been calculated using this modified version of the Keating potential.

Rücker and Methfessel (1995) worked out a generalization of the Keating model for group IV semiconductors and their alloys. First, the energy of mixing relative to the pure materials is separated into two terms, namely $U_{chemical}$ and U_{strain}. The chemical term arises because the strength of an A–B bond is different from the average of the A–A and B–B bond strengths. Secondly, higher-order anharmonic terms are included in order to describe strongly distorted systems. Even for group III–V alloys ($Al_xGa_{1-x}As$ alloys), a modified Keating potential has been presented in the literature (Sim *et al.*, 2005). The anharmonic model potential consists of Coulomb interaction terms in the formulas because of the charge transfer between different types of atoms.

Stillinger–Weber potential

Generally, any empirical potential function could be described by one-, two-, three-body, etc., contributions. If there is no external force, this expansion begins with the pair-wise terms. For sigma-bonded materials the three-body interaction term at least is required. The Stillinger–Weber empirical potential (Stillinger and Weber, 1985) that has been developed for liquid and crystalline phases of silicon is based on two- and three-body interactions:

$$U = A\varepsilon \left\{ \sum_{ij} v_{ij}^{(2)} \left(r_{ij} \right) + \frac{\lambda}{A} \sum_{jik} v_{jik}^{(3)} \left(r_{ij}, r_{ik} \right) \right\}, \tag{3.19}$$

where $\varepsilon = 2.16826\,\text{eV}$. The summations include all pairs *ij* and all triples *jik* of atoms in the system. For $r_{ij}/\sigma < a$, the two-body term is chosen to be

$$v_{ij}^{(2)}\left(r_{ij}\right) = \left\{B\left(\frac{r_{ij}}{\sigma}\right)^{-p} - \left(\frac{r_{ij}}{\sigma}\right)^{-q}\right\}\exp\left(\frac{r_{ij}}{\sigma} - a\right)^{-1}, \qquad (3.20)$$

otherwise $v_{ij}^{(2)}(r_{ij}) = 0$. This definition automatically cuts off at $r_{ij}/\sigma = a$ without discontinuities in any r_{ij}/σ derivatives ($\sigma = 0.20951$ nm). If r_{ij}/σ and r_{ik}/σ are smaller than the cut-off distance a, the three-body part is given by

$$v_{jik}^{(3)}\left(r_{ij}, r_{ik}\right) = \exp\left\{\gamma\left[\left(\frac{r_{ij}}{\sigma} - 1\right)^{-1} + \left(\frac{r_{ik}}{\sigma} - 1\right)^{-1}\right]\right\}\left(\cos\theta_{jik} + 1/3\right)^2, \qquad (3.21)$$

otherwise $v_{jik}^{(3)}(r_{ij}, r_{ik}) = 0$. In practice, only a small subset of these fragments should actually be computed because atoms interact only when their separation is less thanσa. The most "satisfactory" parameter set is the following:

$$A = 7.049556277, \quad B = 0.6022245584, \quad p = 4, \quad q = 0,$$
$$a = 1.80, \quad \lambda = 21.0, \quad \text{and} \quad \gamma = 1.20.$$

In its original form, the Stillinger–Weber potential is not expected to describe accurately the amorphous phase. Parameters were fitted to the crystalline and liquid phases of silicon. A modified Stillinger–Weber potential was fitted directly to the amorphous phase of silicon by Vink *et al.* (2001). As it turned out, the strength of the three-body interaction was boosted by approximately 50% to describe this phase correctly. The improved parameters are $\varepsilon = 1.64833\,\text{eV}$ and $\lambda = 31.5$. The rest stay the same. The modified Stillinger–Weber potential provides a better fit to the measured vibrational density of states spectra. Recently a new parameterization of this potential has been presented for defects and plasticity of silicon base materials (Pizzagalli *et al.*, 2013).

Ding and Andersen (1986) derived, with the help of the Stillinger–Weber potential, a set of parameters to describe germanium. In practice, only three of the parameters are different from those for silicon, for example $\varepsilon = 1.93\,\text{eV}$, $\sigma = 0.2181$ nm, and $\lambda = 31$. The potential provides a good structural representation of a-Ge and c-Ge. Furthermore, it gives good results for phonon dispersion relations of the crystalline phase. Jian, Kaiming, and Xide (1990) presented new sets of parameters for Si and Ge which provide considerable improvements for the phonon dispersion relations.

Although the Stillinger–Weber potential has more flexibility to describe different configurations than the (modified) Keating potential, its transferability to a number of different structures remains in question. To address this issue, a possible improved version of the potential was suggested by Justo *et al.* (1998). Their environment-dependent interatomic potential contains similar terms to those found in the Stillinger–Weber formalism, for example two- and three-body terms, which also depend on the local environment of atom i through its effective coordination number. The effective coordination number is defined by

$$Z_i = \sum_{m \neq i} f(R_{im}), \tag{3.22}$$

where $f(R_{im})$ is a cut-off function that measures the contribution of a neighbor m to the coordination of an atom i in terms of the separation R_{im}; $f(R_{im})$ equals unity when the atomic distance is lower than, or equal to, the first-neighbor distance, and it gently drops to zero between the first- and second-neighbor distances. This leads to a considerable improvement in describing the local bonding for bulk defects and disordered phases of silicon.

Tersoff potentials

A large family of empirical potentials for silicon, carbon, and multicomponent covalent systems was developed by J. Tersoff (1986, 1988a, 1988b, 1988c). Sometimes they are named Abell–Tersoff potentials in order to recall the pioneering work on empirical pseudopotential theory of G.C. Abell (1985). The pair potential

$$U = \frac{1}{2} \sum_{i,j \neq i} f(r_{ij}) \left[A\exp\left(-\lambda_1 r_{ij}\right) - B_{ij}\exp(-\lambda_2 r_{ij}) \right], \tag{3.23}$$

where r_{ij} is the distance between atoms i and j, and A, B, λ_1, and λ_2 are all positive; $f(r_{ij})$ is a cut-off function to restrict the range of the potential. The first term is repulsive, while the second is interpreted as representing the bonding in a Morse-type potential. The parameter B_{ij} implicitly includes the bond order and depends upon local environments. Deviations from the simple pair potential are ascribed to the dependence of B_{ij} upon the local environment. The bonding strength B_{ij} should be a monotonically decreasing function of the number of competing bonds, the strength of the bonds, and the cosines of the angles with the competing bonds.

Optimized Tersoff potential parameters have been derived by Powell, Migliorato, and Cullis (2007) for group III–V zinc blend semiconductors such as GaN, AlN, InN, GaAs, InAs, AlAs, GaP, InP, AlP, GaSb, InSb, and AlSb. Mathematical terms contain 64 fitting parameters. This potential was used mostly for fullerene, nanotubes, and graphenes.

Oligschleger potential

The other important class of amorphous semiconductors is the chalcogenides. The model element for these materials is selenium. An empirical three-body potential has been developed by Oligschleger *et al.* (1996) with the intent of providing both a realistic and a simple description of a selenium–selenium interaction. The most common selenium crystals are built up by either selenium rings or infinite helical chains; in both structures every selenium atom has two nearest neighbors, and this unique property has to be reproduced by the interatomic potential. The authors have taken into account both small selenium clusters and crystalline phases during the parameter fitting of the potential. Its analytical form is given by

$$U = \sum_{i<j} V_2\left(r_{ij}\right) + \sum_{i<j<k} h\left(r_{ij}, r_{kj}, \theta_{ijk}\right) + cyclic\ permutations. \tag{3.24}$$

The three-particle energy is given by

$$h\left(r_{ij}, r_{kj}, \theta_{ijk}\right) = V_3\left(r\right) V_3\left(s\right)\left[b_1(\cos\theta - \cos\beta_2)^2 + b_3 - 0.5 b_1 \cos^4\theta\right], \tag{3.25}$$

with $b_1 = 534.4866$, $b_3 = 511.9572$, and $\beta_2 = 95.3688°$. This three-body term is purely repulsive. Note that $V_{2,3}(r)$ has three different terms in the function of r. If r is lower than 1.6,

$$V_{2,3}(r) = a_{2,3}\exp\left(\alpha r\right) + b_{2,3}\exp\left(\beta r\right) + c_{2,3}\exp\left(\gamma r\right). \tag{3.26}$$

For $1.6 < r < r_{2,3}$,

$$V_{2,3}(r) = d_{2,3}(r - r_{2,3})^5 + e_{2,3}(r - r_{2,3})^4 + f_{2,3}(r - r_{2,3})^3. \tag{3.27}$$

Otherwise, $V_{2,3}\left(r\right) = 0$. Note that in Table I in the original paper (Oligschleger *et al.* (1996), p. 6165) the parameter $a = 9281.2$ in the two-body case. The other misprint is that $\alpha = -7.9284$ instead of -7.984. Correct parameters in reduced units (see the original paper) can be taken from Table 3.2.

Table 3.2. *Revised parameters for selenium potential.*

	Two-body case	Three-body case
a	9281.2	8.8297
b	0.26802	−2.5932
c	−16.599	−6.9384
α	−7.9284	−0.47601
β	−0.000077781	−1.5637
γ	−1.8634	−0.37049
d	1.86825	0.22556
e	4.58628	0.12527
f	3.62029	−0.21019
$r_{2,3}$	2.37	2.35

Altogether, the 23 parameters are fitted to density-functional calculations. The potential has been carefully tested and compared to experimental results. The descriptions of atomic geometries and bonding energies are excellent, but a weak point of the potential lies in the quantitative description of vibration properties. However, the overall performance of the potential is outstanding when considering its effectiveness and simplicity.

Tight-binding models

Tight-binding (TB) models offer a semiempirical quantum chemical approach to describing the interaction between atoms. The TB model is useful in generating and describing the atomic structure of amorphous materials. We focus only on orthogonal TB models that have been developed for carbon, silicon, and selenium. The total energy of the system is written as follows:

$$E_{tot} = E_b + E_{rep}, \tag{3.28}$$

where E_b is the sum of electronic eigenvalues ε_i over all occupied electronic states, given by

$$E_b = 2\sum_{i}^{n_{occ}} \varepsilon_i, \tag{3.29}$$

and E_{rep} is the short-range repulsive energy. The electronic eigenvalues are obtained by diagonalizing an empirical tight-binding Hamiltonian. The on-site elements are the atomic orbital energies of the corresponding atoms, whereas the off-diagonal elements of the tight-binding Hamiltonian are described by a set of orthogonal sp^3 two-center hopping parameters, $Vss\sigma$, $Vsp\sigma$, $Vpp\sigma$, and $Vpp\pi$, scaled with interatomic separation r as a function $s(r)$. The remainder of the E_{tot} is modeled by a short-range repulsive term E_{rep} given by

$$E_{rep} = \sum_i f\left(\sum_j \Phi(r_{ij})\right), \tag{3.30}$$

where $\Phi(r_{ij})$ is a pair-wise potential between atoms i and j, and f is expressed as a fourth-order polynomial.

In order to model the amorphous structure, the TB model has to be transferable. This can be achieved by adopting a suitable functional form, in this case the one suggested by Goodwin, Skinner, and Pettifor (1989). Thus the scaling functions $s(r)$ and $\Phi(r)$ are given by

$$s\,(r) = (r_0/r)^n \exp\left\{n\left[-\left(\frac{r}{r_c}\right)^{n_c} + \left(\frac{r_0}{r_c}\right)^{n_c}\right]\right\} \tag{3.31}$$

and

$$\Phi\,(r) = \Phi_0(d_0/r)^m \exp\left\{m\left[-\left(\frac{r}{d_c}\right)^{m_c} + \left(\frac{d_0}{d_c}\right)^{m_c}\right]\right\}, \tag{3.32}$$

where r_0 denotes the nearest-neighbor atomic separation in the diamond structure. Parameters for carbon are found in Xu *et al.* (1992) and for silicon by Kwon *et al.* (1994). A similar formalism has been used for selenium by Molina, Lomba, and Kahl (1999).

More sophisticated TB models are also available; see, for example, Tang *et al.* (1996) for carbon, and Wang, Pan, and Ho (1999) for silicon.

A more accurate approach is to derive the atomic interaction directly from the electronic ground state, where the total energy functional is calculated using density-functional theory. The difficulty of DFT is in obtaining an accurate exchange and correlation interaction inside a many-body system. Nowadays this is a rapidly developing and most popular method in solid state physics. The computer cost is much higher than that for TB models, and it is comparable to the Hartree–Fock *ab initio* method, which is not the best for structure investigation.

Computer simulation methods

During the past decades research has focused on the construction of realistic atomic configurations for amorphous materials using computer simulations. Computations should not substitute for lack of knowledge of atomic-scale configuration, but instead be a useful tool for constructing atomic-structure models. There are two main possibilities for structural modeling methods of disordered configurations at the atomic level: a stochastic method (Monte Carlo (MC) type) and a deterministic method (molecular-dynamics (MD) simulations).

Monte Carlo methods

The name Monte Carlo arises from the famous casino in Monaco. The method involves the use of random numbers to govern atomic displacements during computer simulation processes. Several algorithms fall under the MC method: traditional MC, reverse MC (McGreevy and Pusztai, 1988), quantum MC, kinetic MC, path integral MC methods (Herrero, 2000), etc. In this section, only the first two techniques are discussed. Systems must have a large number of degrees of freedom, which are investigated by the MC method. The traditional MC system generating a three-dimensional particle configuration of amorphous materials searches for a local minimum of total energy on the energy hypersurface. In order to calculate the energy, we need to describe the influence on each atom from the atoms that surround it. Two different interactions are usually applied for relaxation: one is known as the empirical potentials and the other is based on different quantum-mechanical approaches, such as tight-binding models, density-functional theory, or Hartree–Fock approximations.

The initial configuration could be any atomic arrangement (even a distorted crystalline structure), although the perfect crystal provides the absolute minimum total energy on the energy hypersurface. A randomly chosen atom is displaced to a new position that is determined with the help of a random-number generator. In most of the simulations, a maximum displacement is applied as a commonly used constraint. The energy variation is a crucial parameter for the MC method. If a MC step provides a downhill motion on the energy hypersurface, then the new configuration is accepted, and this reordered structure becomes the initial atomic arrangement for the next MC step. In an uphill case, the new position is rejected only conditionally. This decision is the essential part of MC simulations.

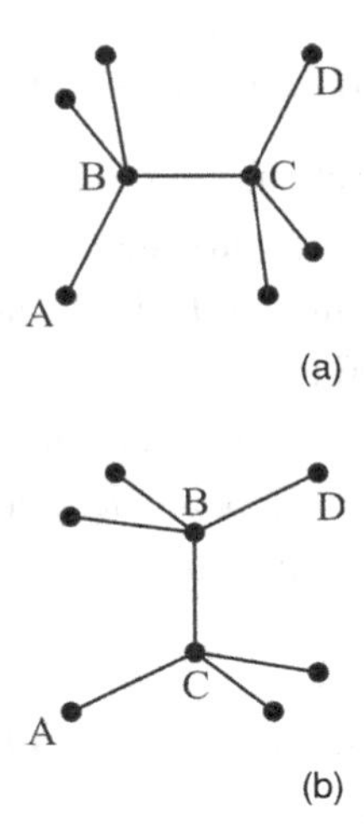

Figure 3.10. Bond transposition process for a-Si (a-Ge) structural model constructed by Wooten *et al.* (1985) using MC simulation. (a) Configuration of bonds in diamond structure; (b) modified configuration after bond transposition.

A simple explanation of the selection rule is the following. If uphill motion is always rejected, then the procedure reaches the first shallow energy minimum and there is no further possibility to search for a deeper minimum which belongs to a more realistic non-crystalline atomic arrangement. A precise description of MC philosophy can be learned from any elementary statistical physics book.

Amorphous silicon model constructed by Monte Carlo simulation

The first state-of-the-art traditional MC simulation was performed for pure a-Si and a-Ge by Wooten *et al.* (1985). The main purpose was to construct a homogeneous structure without any defects, i.e. Si atoms have four nearest neighbors and there are no voids inside the model structure. The starting atomic configuration was a simple cubic diamond structure, and a number of bond transpositions were carried out (see Figure 3.10). This process destroys the translation and topological symmetries and includes fivefold and sevenfold rings (Wooten and Weaire, 1984). Similar bond transposition in two-dimensional carbon systems (fullerenes, nanotubes, and graphenes) is called the Stone–Wales transformation (Stone and Wales, 1986). After randomization using the Metropolis MC algorithm (Metropolis *et al.*, 1953), the structure was relaxed by the Keating empirical potential. A continuous random network with 216 atoms has periodic boundary conditions in three dimensions in order to preserve local order on the surface. The topology includes fivefold, sixfold, sevenfold, and eightfold rings. In the diamond structure, only sixfold and eightfold rings should be found. Based on a close analogy between silicon and germanium or saturated carbon compounds, this model can be rescaled by nearest-neighbor distances and can be applied for a-Si, a-Ge, and tetrahedral a-C (ta-C) configurations. Since then, several computer-generated models have been constructed using various classical empirical potentials (Kelires and Tersoff, 1988; Ishimaru, Munetoh, and Motooka, 1997; Ishimaru, 2001; Vink *et al.*, 2001) or by applying different quantum-mechanical methods (Drabold *et al.*, 1991; Stich, Car, and Parinello, 1991; Tóth and Náray-Szabó, 1994; Hensel *et al.*, 1996; Yang and Singh, 1998; Cooper, Goringe, and McKenzie, 2000, Valladares *et al.*, 2001). Despite this, the WWW model developed by Wooten *et al.* (1985) is still considered to be the best defect-free three-dimensional atomic-scale representation of covalently bonded a-Si (and a-Ge). The WWW or modified WWW models were applied for the electronic structure calculations, and the results from these works suggest that the covalently bonded a-Si structures are not completely disordered. The bonds between atoms and the coordination numbers are similar to those in the crystalline phase.

A special MC simulation technique, in which the energy is calculated by quantum-mechanical methods, was proposed by Tóth and Náray-Szabó (1994). The applied semiempirical fragment self-consistent field technique divides the periodic simulation cell into two parts. The first is the subsystem where the important change (the random motions) occurs, and the rest comprises the environment exerting only secondary effects on the former part. The conventional self-consistent field equations have to be solved only for the critical subsystem. In this way, the computational efforts are decreased drastically, as the dependence on the number of atoms in the environment reduces to quadratic instead of cubic or quartic, as in conventional semiempirical or *ab initio* methods. This quantum MC technique was performed first for a sigma-bonded a-Si structure investigation (Tóth and Náray-Szabó, 1994).

Reverse Monte Carlo methods

Another version of the MC method – the reverse Monte Carlo (RMC) simulation – has been developed by McGreevy and Pusztai (1988) for the structure investigation of disordered condensed phases and liquids. Initial results for liquid argon were presented by the authors. This inverse problem solution method is convenient for the analysis of diffraction data and at the same time for modeling the amorphous structures. The RMC simulation is based on the results of diffraction measurements and is free from the description of the interaction between atoms. This is the only method generating three-dimensional particle configurations that are consistent with the experimentally derived structure factor or the radial distribution function.

Theoretically, both functions, the structure factor and the RDF, can comprise the initial data for RMC simulation, but use of $S(Q)$ has an important advantage. The main problem that occurs at any diffraction measurement – the measurement of Q in only a limited interval and Fourier transformation in an infinite interval – is bypassed. To speed up the simulation and/or to reach more realistic configurations, constraint(s) could usually be applied. To start the RMC simulations, a given number of particles are confined in a well-defined box. The simulation derives the density that is the most important constraint. Another constraint that is used several times is the coordination number. It can easily be prescribed that particles must strictly have a given number of particles within a fixed distance. A hard core diameter (the lower limit of bond length) can also be applied in order to avoid physically meaningless configurations. For systems where atomic interactions are covalent, i.e. bonds are directed, introducing constraints on bond angles might also be a useful constraint.

The use of constraints modifies the original RMC algorithm in the following way.

(a) Start with an initial particle configuration. Calculate its RDF or structure factor. Calculate also the difference between the simulated and experimental RDFs or structure factors (χ_0^2). The comparison with experiment is quantified using a function of the form $\chi^2 = \sum \frac{(y_{exp} - y_{sim})^2}{\sigma^2}$, where y_{exp} and y_{sim} are the measured and calculated quantities; σ is a measure of the accuracy of the measurement. The sum includes all points in a function such as the RDF or structure factor.
(b) A new, trial, configuration is generated by the random motion of a particle.
(c) Check whether the new configuration satisfies the constraint(s) applied. If not, start again from (b). (This is an additional step to the standard RMC simulation.)
(d) Provided that the constraints are satisfied, calculate χ_n^2 for the new trial configuration.

(e) If $\chi_n^2 < \chi_{(n-1)}^2$, the new configuration becomes the starting configuration, i.e. the move is accepted; otherwise it is accepted with a probability that follows the normal distribution.
(f) Repeat the process from (b) until χ_n^2 converges to its "equilibrium" value.

In this way, sets of particle coordinates are generated which are consistent with a given diffraction data set and with the constraint(s) applied. Figure 3.11 displays three stages of a RMC calculation in two dimensions. Figure 3.11(a) shows the initial crystalline configuration (right), and the measured (thick solid line) and calculated (thin solid line) pair correlation functions (left). Figure 3.11(b) displays a middle stage of the RMC simulation, and Figure 3.11(c) represents a stage where the fit is near optimum (L. Pusztai, private communication). Free RMC computer code is available at http://www.szfki.hu/~nphys/rmc++/opening.html. A brief description of the computer code can be obtained from Gereben *et al.* (2007) and Gereben and Pusztai (2012).

Constraints

In order to derive constraints, the following question should be answered: what natural local configurations can be found in nature? During prepartion of the amorphous phase, the covalently bonded atoms tend to seek out one of the different local configurations. Information about these local orders of the amorph–ous structures might be provided by analyzing the embedded fragments inside different large molecules. The environment might be considered as a "white noise" around the fragments. The Cambridge Structural Database (CSD) (Allen, Kennard, and Taylor, 1983), the world's largest experimentally determined crystal structure, has within it the results of x-ray and neutron-diffraction studies. The CSD is designed as a critically evaluated numerical resource, containing 3D atomic coordinates. It provides a good opportunity to obtain measured lower and upper limits for bond lengths, bond angles, second-neighbor distances, etc., if the CSD contains a large enough number of diffraction records.

A systematic search of structural data can be carried out for X–X and X–X–X fragments on this database, where X denotes the element forming the disordered system. We have collected the experimentally determined structural data of molecules containing Si–Si–Si fragments (Kugler and Varallyay, 2001) and for Se–Se–Se (Hegedus and Kugler, 2005). Recently, a new search was made[1],

[1] Lukács, R. and Harmath, V. (2011), private communication.

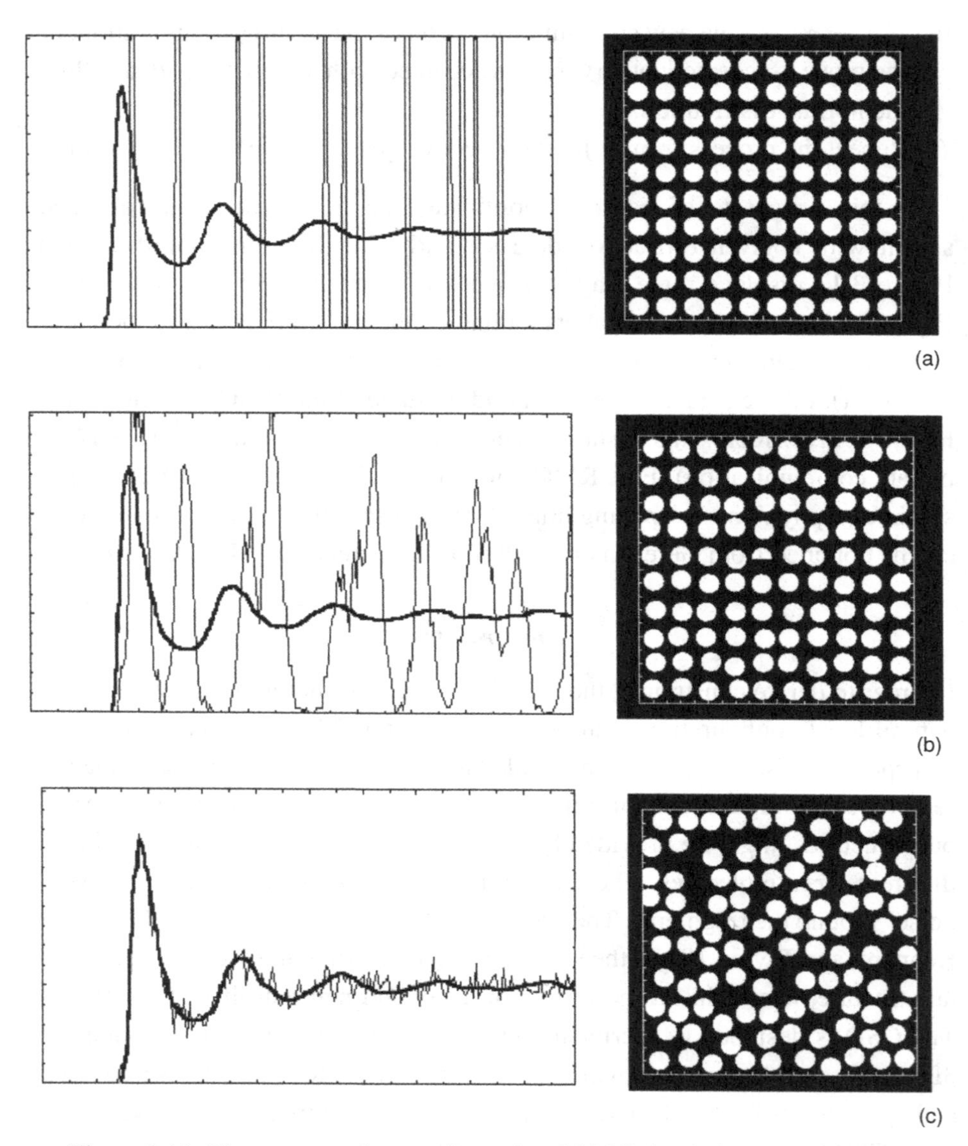

Figure 3.11. Three stages of a two-dimensional RMC simulation. (Provided by L. Pusztai (Budapest).)

and Figures 3.12 and 3.13 display the results of these bond-angle distributions for a-Si and a-Se, respectively. The number of Si–Si–Si fragments is equal to 8071, whereas for Se–Se–Se it is less (905 fragments). The results of a search for a-Si are surprising. The overwhelming majority of these points fall in the expected region, i.e. around 0.235 nm and 109.47°. The minimum bond length

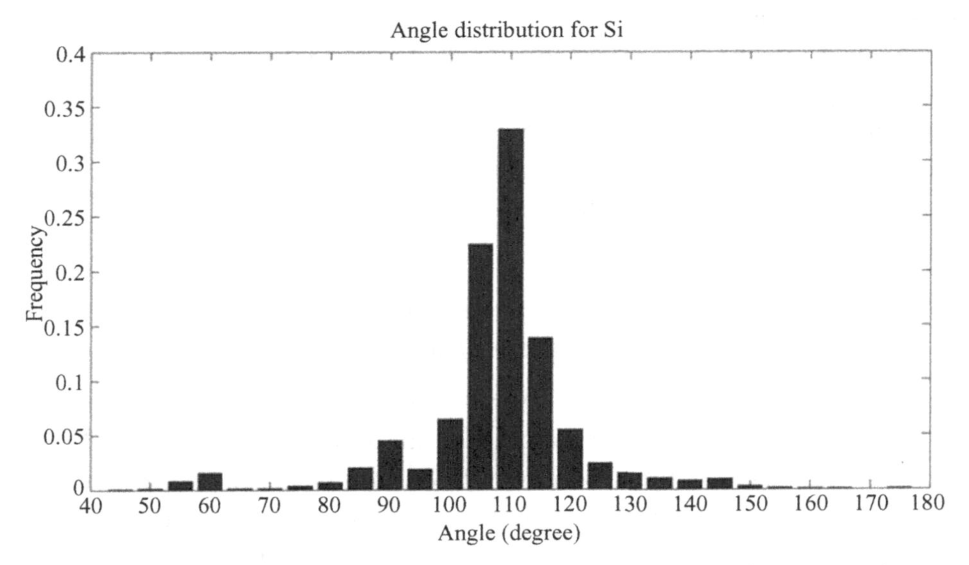

Figure 3.12. Histogram of bond-angle distribution of Si–Si–Si fragments collated from the CSD.

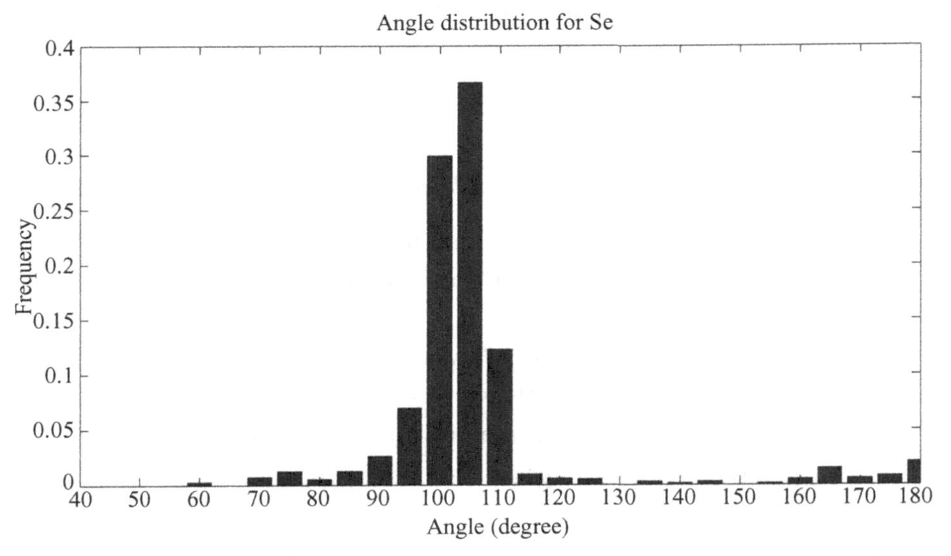

Figure 3.13. Histogram of bond-angle distribution of Se–Se–Se fragments collated from the CSD.

is about 0.22 nm, whereas the maximum is 0.27 nm. The average bond length and angle are 0.237 nm and 106.3°, respectively. There are some extrema. Two well-defined, unexpected regions can also be found, for example there are some significant angles in an interval of 75°–96°. These angles belong to the near planar square arrangement. Several squares were found in the database. Bond angles around 60° can be also found. The second conclusion of bond-angle analysis is that nearly equilateral triangles of Si_3 are present among the fragments. This is the other unexpected region, and most of the a-Si theoretical models do not contain this type of structure, i.e. three- and four-membered rings. A similar conclusion can be delivered for a-Se.

Tetrahedral amorphous carbon model constructed by reverse Monte Carlo simulations

Tetrahedral a-C (ta-C) is one of several forms where most atomic sites are fourfold coordinated. A neutron-diffraction measurement by Gilkes and co-workers (Gilkes, Gaskell, and Robertson, 1995) has concluded that the structure of ta-C consists predominantly of a disordered tetrahedral network, as does a-Si; a-C is more complicated because a variety of local arrangements exists. In this case, 3000 carbon atoms were put into a box with edges of 2.714 nm (Walters *et al.*, 1998). Four models were constructed with different constraints. Particular attention was paid to local bonding environments. Walters *et al.* concluded that better agreement with the measured data is achieved if the model is allowed to include threefold and fourfold rings.

Amorphous silicon model constructed by reverse Monte Carlo simulation

The first RMC simulations including constraints were performed for pure a-Si by Kugler *et al.* (1993a). A neutron-diffraction measurement on a pure evaporated a-Si sample was performed at the high flux reactor in the ILL, Grenoble. The experimentally measured structure factor was used as the input data for the RMC calculations. Three structural models were constructed. In all the simulations $N = 1728$ particles were confined in a cubic box of sides $L = 3.2$ nm. This set-up (constraint) yielded the experimental microscopic number density $\rho = 0.0505$ Å^{-3}. A hard core diameter (lowest limit of bond length) of 0.22 nm was also applied in order to efficiently avoid physically meaningless configurations. About one million accepted steps were completed during these three different runs.

The first simulation (referred to as Model 1) is the result of the RMC without any other constraint. Model 2 was produced using a constraint on the coordination number: every atom in the configuration was required to have exactly four

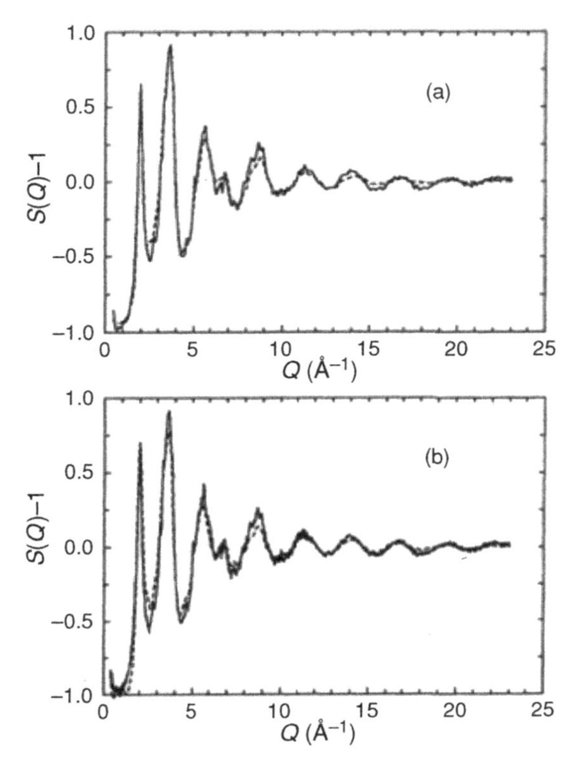

Figure 3.14. (a) Comparison of the RMC solution for Model1 and the experimental $S(Q)$ for a-Si. (b) The same comparison for Model3 for a-Si. (Taken with permission from Kugler *et al.* (1993). *Phys. Rev. B*, **48**, 7685. Copyright 2013 by the American Physical Society.)

neighbors in the first coordination shell. The upper limit for the bond length was 0.27 nm. The $g(r)$ function obtained by the Fourier transformation of the measured $S(Q)$ almost reached its minimum between the first and second main peaks at 0.27 nm. Furthermore, the CSD suggests a similar value for the maximum first-neighbor distance. Model 3 satisfied the most complicated constraint. In addition to the atoms having four neighbors, if an attempted move resulted in newly formed bond angles with an average that did not fit into a normal-like distribution centered on the tetrahedral angle, the move was immediately rejected. In systems such as a-Si, it is a normal requirement that most of the angles should be roughly tetrahedral. Angles far from the tetrahedral value are less probable, but permitted. Figure 3.14(a) shows the comparison of the RMC solution for Model 1 and the experimental $S(Q)$, and Figure 3.14(b) gives the same comparison for Model 3. The overall agreement is quite good. It is thought

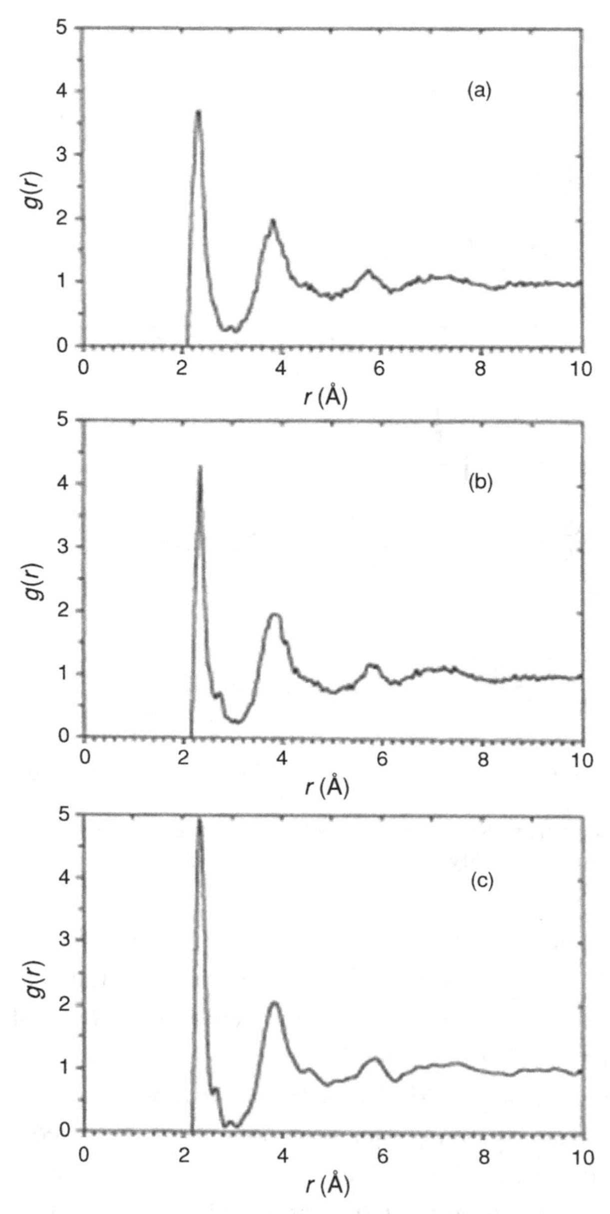

Figure 3.15. (a), (b) Pair correlation functions calculated using RMC with different constraints. (c) Pair correlation function obtained by Fourier transformation of measured $S(Q)$. (Taken with permission from Kugler *et al.* (1993). *Phys. Rev. B*, **48**, 7685–7688. Copyright 2013 by the American Physical Society.)

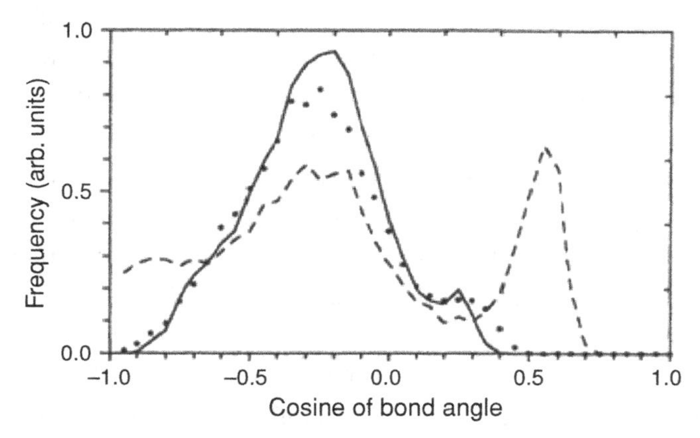

Figure 3.16. Cosine of bond-angle distributions for three different models. (Taken with permission from Kugler *et al.* (1993). *Phys. Rev. B*, **48**, 7685–7688. Copyright 2013 by the American Physical Society.)

that the greater amplitude of the oscillations at higher Q-values and the small-angle-scattering part of the $S(Q)$ curve could be fitted more successfully using a much bigger box.

The RDF for Model 1 (identical to Model 2) is displayed in Figure 3.15(a), and that for Model 3 is given in Figure 3.15(b). Figure 3.15(c) depicts $g(r)$ obtained by using a Fourier transform of the measured $S(Q)$. The characteristics of these functions are rather similar. Cosine distributions of bond angles ($\cos\theta$) lead to significant differences between microscopic structures obtained using three different constraints (Figure 3.16). This makes clear that a diffraction experiment by itself provides insufficient information to determine unambiguously the microscopic local structure. Identical $S(Q)$ and/or $g(r)$ functions can be consistent with substantially different bond-angle distributions. The cosine distribution for Model 1 has a large peak at around 60°, and also a large proportion of angles were found near 180°. However, an intense peak around 109.5° is provided by the model. These results are completely different from those of the other two models. Model 2 reflects probably the most fundamental requirement that atoms in any solid form of Si should be fourfold coordinated. This simple constraint greatly reduces the freedom of moving atoms around. The main problem with this picture might be the probable over-representation of angles higher than 130°. Model 3 was introduced to narrow the cosine distribution further, but no essential narrowing could be achieved compared with Model 2.

Wooten's model of a-Si contains internal bond angles between about 90° and 150°. The classical empirical Keating potential was applied for the construction in which the interaction has a quadratic energy term for the difference between the cosine bond angle and the cosine ideal bond angle. This term avoids the large deviation from the canonical value of bond angle. Note that the RMC simulation does not take the energy into account. The existence of angles smaller than 90° or 60° also agrees with the CSD. A Japanese group investigating a-Si:H by RMC simulation also concluded the existence of small bond angles (Tabuchi *et al.*, 2004). Furthermore, another Japanese group (Ishimaru, Yamaguchi, and Hirotsu, 2004) investigated silicon–germanium alloys by a MD method using a classical empirical potential of Tersoff (Tersoff, 1989), and they also observed bond angles of 60°.

A comparison between the experimental structure factors of evaporated and ion-implanted a-Si has been modeled by the RMC simulation (Pusztai and Kugler, 2005). A detailed comparison, in terms of the pair correlation function and the cosine distribution of bond angles, has been reported for the two materials. It is found that for an acceptable RMC reproduction of the measured structure factors, the evaporated models must contain more "small" bond angles than their necessary abundance in the implanted models.

RMC modeling for comparison of a-C, a-Si, and a-Ge structures

A series of RMC simulations were carried out for a-C, a-Si, and a-Ge (Gereben and Pusztai, 1994). A number of constraints were applied during the calculations to test how many different assumptions are consistent with the given experimental data sets. The results of the unconstrained calculations seem to be the best fit to these samples, but these models are not satisfactory from the electronic structure point of view. The basic constraint that causes a problem is the exact coordination number of four. The most characteristic common feature of these amorphous semiconductors is not fourfold coordination.

Amorphous selenium model constructed by reverse Monte Carlo simulation

Similar RMC work was carried out on the model material (a-Se) of chalcogenide glasses by Jóvári and co-workers (Jóvári *et al.*, 2003). Neutron-diffraction measurements were performed on the SLAD instrument at NFL, Studsvik, on a

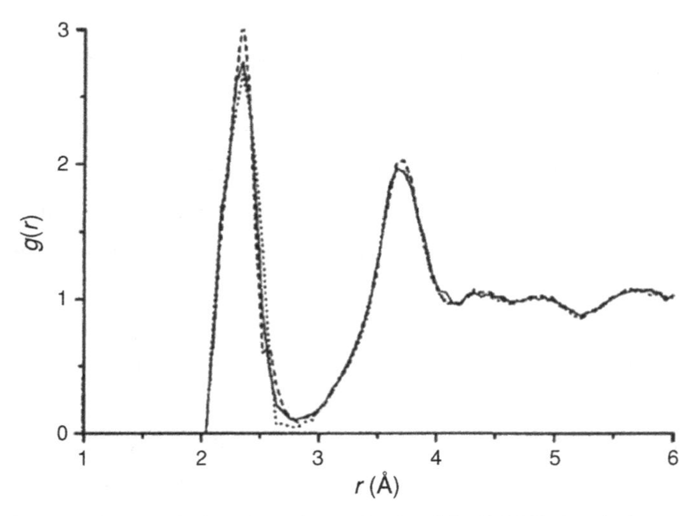

Figure 3.17. Pair correlation functions obtained by RMC simulation applying different constraints. Solid line: hard-sphere constraints only; dashed line: long chains; and dots: Se_8 rings. (Taken with permission from Jóvári, Delaplane, and Pusztai (2003). *Phys. Rev. B*, **67**, 172201. Copyright 2013 by the American Physical Society.)

ball-milled a-Se sample. The obtained structure factor was interpreted by RMC simulation. Selenium has several crystalline forms, consisting of Se_8 rings or chains. The authors concluded from the diffraction data that it is not possible to determine whether a-Se consists of long chains, Se_8 rings, or a mixture of both phases. It is possible to construct both of the models whenever the structure factors agreed quantitatively with measured data (Figure 3.17). The measured sample was made from crystalline Se consisting of long chains; it is suggested that the investigated sample contained primarily chain-like structures.

Molecular-dynamics simulation

An alternative 3D structure model construction is the deterministic molecular-dynamics (MD) method for describing non-crystalline systems. The primary application of MD is to follow the preparation process *in situ* and to reach a condensed phase. In this process, all that is assumed is the validity of classical mechanics. This approximation is never truly invalid. An indicator of the validity of classical mechanics is the de Broglie wavelength, $\lambda = h/\sqrt{2mE}$. For a particle moving with a kinetic energy of 1 eV, the de Broglie wavelength of the particl is 1.2 nm (if it is an electron), 0.3 nm (hydrogen), 0.1 nm (carbon), and 0.005 nm (silicon). In the condensed phase the characteristic distance is the interatomic spacing, which is usually equal to 0.1–0.3 nm. The motion of a 1 eV electron in the solid state therefore can only be described by including the quantum effect in the dynamics. On the other hand, classical mechanics is a good approach to take when considering carbon, silicon, or selenium, etc. The atoms are treated as point particles because the de Broglie wavelengths are small enough.

The Newtonian equations of motion can be solved if the position and velocity of N particles are given at a time t. It should be mentioned that an important conclusion of quantum mechanics, the Heisenberg uncertainty relation, does not allow this statement to remain valid. The uncertainty relation forbids knowing the exact atomic position and momentum simultaneously with arbitrary precision ($\Delta p_x \Delta x \geq \hbar/2$).

The simplest way of integrating the MD equations of motion is to use the Verlet algorithm (Verlet, 1967). The atomic motion can be followed by the MD method. The basic idea is to construct third-order Taylor expansions for $\boldsymbol{r}(t + \mathrm{d}t)$ and $\boldsymbol{r}(t - \mathrm{d}t)$, where $\boldsymbol{r}(t)$ is the atomic position at time t:

$$\boldsymbol{r}(t + dt) = \boldsymbol{r}(t) + \boldsymbol{v}(t)\,dt + \frac{1}{2}\boldsymbol{a}(t)dt^2 + \frac{1}{6}\boldsymbol{b}(t)\,dt^3 + O(dt^4), \quad (3.33)$$

$$\boldsymbol{r}(t - dt) = \boldsymbol{r}(t) - \boldsymbol{v}(t)\,dt + \frac{1}{2}\boldsymbol{a}(t)dt^2 - \frac{1}{6}\boldsymbol{b}(t)\,dt^3 + O(dt^4). \quad (3.34)$$

Adding the two expressions gives

$$\boldsymbol{r}(t + dt) = 2\boldsymbol{r}(t) - \boldsymbol{r}(t - dt) + \boldsymbol{a}(t)\,dt^2 + O(dt^4), \quad (3.35)$$

and then we get

$$\boldsymbol{v}(t) = \frac{\boldsymbol{r}(t + dt) - \boldsymbol{r}(t - dt)}{2dt} + O(dt^3). \quad (3.36)$$

Acceleration $\boldsymbol{a}(t)$ at time t can be calculated from the atomic interactions as follows:

$$\boldsymbol{a}(t) = -(1/m)\mathrm{grad}\,V(\boldsymbol{r}(t)). \quad (3.37)$$

The application of the Hellmann–Feynman theorem is required to calculate the forces in the quantum-mechanical case (Hellmann, 1933; Feynman, 1939). Equation (3.37) is the basic equation of the Verlet algorithm. On the basis of these equations, therefore, knowing (i) the positions of each particle at time t

and at the earlier time $(t - dt)$, and (ii) $V(r)$, we can derive the new configuration at time $(t + dt)$. Knowledge of $\boldsymbol{r}(t - dt)$ provides a little difficulty at the first MD step, but this is easy to handle as long as we know the initial velocity. Furthermore, we can calculate the atomic velocities at time t, which is important for the temperature estimation. The local error in a given position is proportional to $O(dt^4)$. If we calculate the cumulative error for the positions over a given time period T, we obtain

$$error\,(T) = O(dt^2). \tag{3.38}$$

More commonly, a velocity Verlet algorithm is used (Swope *et al.*, 1982). This new method uses a similar approach but explicitly incorporates velocity, solving the first time step problem in the basic Verlet algorithm. The starting equations are

$$\boldsymbol{r}\,(t + dt) = \boldsymbol{r}\,(t) + \boldsymbol{v}\,(t)\,dt + \frac{1}{2}\boldsymbol{a}\,(t)\,dt^2 \tag{3.39}$$

and

$$\boldsymbol{v}\,(t + dt) = \boldsymbol{v}\,(t) + \frac{\boldsymbol{a}\,(t) + \boldsymbol{a}(t + dt)}{2}dt. \tag{3.40}$$

First, $\boldsymbol{r}\,(t + dt)$ must be calculated; after that one should compute $\boldsymbol{a}(t + dt)$ using the new position $\boldsymbol{r}(t + dt)$, and finally $\boldsymbol{v}\,(t + dt)$ must be derived. The error for the velocity Verlet is of similar order to that for the basic Verlet algorithm. The same computer memory is needed for this modified version.

In the leapfrog integration (similar to the velocity Verlet method), the velocity is calculated in two steps. The new positions are calculated as in the velocity Verlet algorithm. Then, the velocities at a half-time step are calculated from the old velocities and accelerations:

$$\boldsymbol{v}\left(t + \frac{1}{2}dt\right) = \boldsymbol{v}\,(t) + \frac{1}{2}\boldsymbol{a}(t)dt. \tag{3.41}$$

Then the $(t + dt)$ terms are computed, and finally the velocity can be obtained in the following way:

$$\boldsymbol{v}\,(t + dt) = \boldsymbol{v}\left(t + \frac{1}{2}\right) + \frac{1}{2}\boldsymbol{a}\,(t + dt)\,dt. \tag{3.42}$$

It has been shown that this leapfrog algorithm is numerically more stable than the version in which the velocities are calculated in one step.

There are other methods for the numerical integration of ordinary differential equations of order 2, such as Beeman's algorithm (Beeman, 1976), or the well-known Runge–Kutta method, but in most cases the velocity Verlet algorithm is applied.

Temperature control

In an amorphous sample, the preparation temperature plays a crucial role. The temperature must be controlled during quenching from the liquid state and in the deposition to a substrate. Most of the main preparation techniques utilize low-temperature-induced amorphization. Each of these preparation techniques is highly non-equilibrium in nature. A heat bath removes energy from the sample during the amorphization. At every MD simulation step the temperature must be monitored. The instantaneous temperature is given by the following well-known statistical mechanics equation:

$$\frac{3}{2}Nk_BT = \sum_{i=1}^{N} \frac{1}{2m} p_i^2. \tag{3.43}$$

To simulate the heat conduction outside the simulation box, and to bring the structure back to the desired temperature, one is required somehow to control the temperature. During the process of film growth, particles bombard the growing film. Deposition to a substrate requires that the substrate temperature is kept constant. The instantaneous kinetic energies of the substrate atoms (and the temperature of the substrate), however, may increase during MD simulations due to particle impact, and they are dependent on the energy of the incoming particle. At rapid cooling, the characteristic rate of quenching is around 10^6 K s^{-1}, or lower in experiments. A variety of thermostat methods used to control the temperature can be found in literature, but the Nosé–Hoover thermostat (Nose, 1984; Hoover, 1985; Chen, Wu, and Cheng, 2011) and the simpler velocity-rescaling methods are the most popular. Equations describing the Nosé–Hoover thermostat have a free parameter, and sometimes its derivation is difficult.

A MD computer code (ATOMDEP) has been developed to simulate the preparation procedure of amorphous semiconductors, which are usually grown by a vapor-deposition technique on a substrate (Kohary and Kugler, 2001). It should be noted that experimentally no rapid quenching preparation method has been reported for group IV amorphous materials. To integrate the MD equations of motion, the velocity Verlet algorithm was included with a time step equal to about 1 fs. The substrate temperature was fixed by velocity rescaling. During the growth procedure, an atom started to move toward the target surface. In order to avoid surface effects, periodic boundary conditions in two dimensions were adopted. The initial x, y coordinates and the deposition energies were randomly distributed. The bottom layer of the substrate was fixed at its ideal lattice site in order to stabilize mechanically the substrate.

A typical deposition rate during sample preparation in the laboratory is around 10^{14} atoms s^{-1} cm^{-2}. The surface of the substrate in computer experiments is around 10^{-14} cm^2, i.e. one incoming atom per second should reach the substrate. This deposition rate presents a difficulty for the simulation, and it is impossible to handle easily using MD with today's computer facilities within a tight-binding atomic interaction model. Despite this disadvantage, the high theoretical deposition rate applied is low enough to warrant the relaxation of the previous atom before the next deposition. This developed computer code was successfully applied to prepare a-C, a-Si, and a-Se, and this will be discussed in the following section.

Amorphous carbon model construction by molecular-dynamics simulation

MD simulations for a-C were carried out to study the dynamics of the growth process (Kohary and Kugler, 2001). The tight-binding Hamiltonian of Xu *et al.* (1992) was used to calculate the interatomic potential between carbon atoms. This realistic TB potential has already been successfully applied for different carbon systems, even for the preparation of fullerenes (Laszlo, 1998, 1999).

The simulation technique took the following form. An ideal diamond film consisting of 120 carbon atoms was employed to model the substrate. The rectangular simulation cell was open along the positive z axis (the [111] direction in our case). Periodic boundary conditions were used in two (x and y) directions. Carbon atoms in the bottom substrate layer were fixed at their ideal lattice, whereas the rest of the 96 atoms were allowed to move with full dynamics. To simulate the constant substrate temperature, the kinetic energy of the movable atoms in the substrate was rescaled at every time step. The time step was chosen to be 0.5 fs.

The bombarding atoms were randomly placed in the x and y directions above the substrate, as near as possible to any other atom, but no closer than the cut-off distance of the potential. The initial velocities of bombarding atoms were directed to the substrate. The initial velocities were chosen using the simple

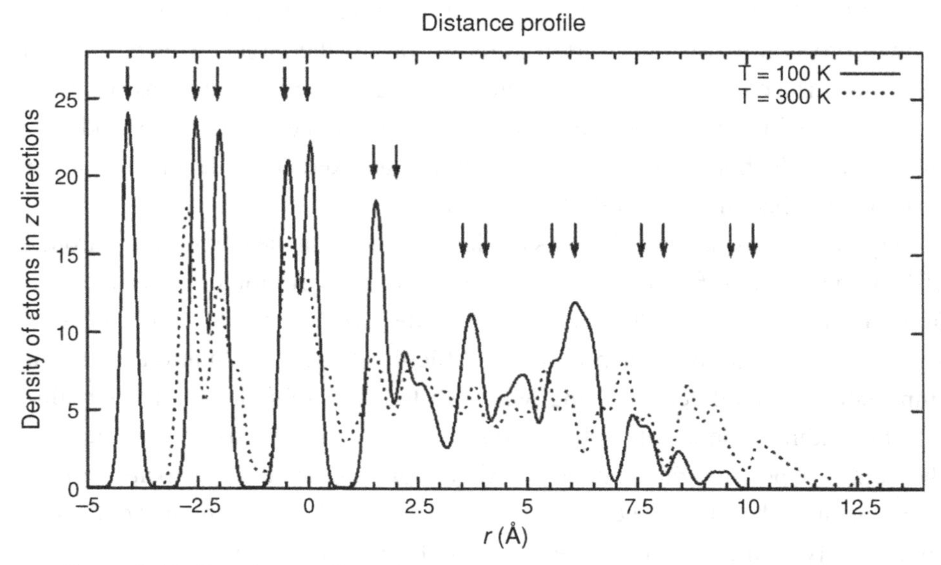

Figure 3.18. Density profiles of two networks (T = 100 K and T = 300 K) are displayed perpendicular to the substrate surface. The arrows represent the layer positions of a perfect diamond crystal. A memory effect is observed near the surface of the substrate. (Taken with permission from Kohary and Kugler (2001). *Phys. Rev. B*, **63,** 193404. Copyright 2013 by the American Physical Society.)

relation $v^2 = (2E/m)(1.2 - 0.4p)$, and the directions were determined using $\theta = 120° + p60°$ and $\varphi = p360°$, where θ and φ are the polar angles, and p is a uniformly distributed random number between 0 and 1. The frequency of the atomic injection was, on average, $f = 1/125$ fs^{-1}. This flux is orders of magnitude greater than the deposition rate usually applied in real experiments. The lower substrate temperatures applied in the simulations result in faster energy dissipation, which compensates for the high deposition rate.

Models were constructed with average bombarding kinetic energies (E) of 1 eV and 5 eV on the substrate with temperatures of 100 K and 300 K. In all cases the structures were relaxed for 5 ps after deposition times of 25 ps and 40 ps. The density profiles of the two networks (T = 100 K and T = 300 K) are displayed perpendicular to the substrate surface in Figure 3.18. The arrows represent the layer positions of a perfect diamond crystal. The difference in temperature between two substrates causes differences in memory effects. The structure on the surface at 100 K has a more pronounced layering effect than the other network on the substrate at room temperature. In the first 0.3–0.4 nm

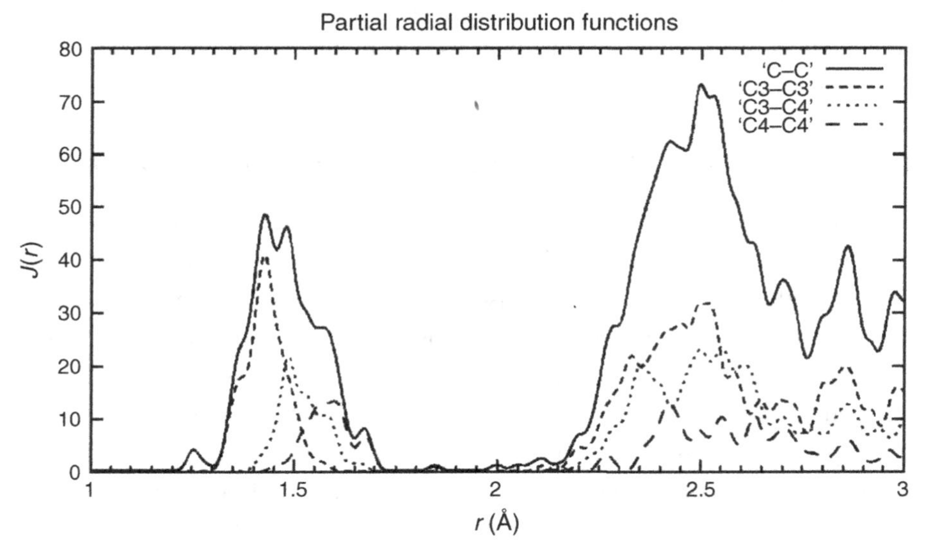

Figure 3.19. Partial radial distribution functions $J(r)$ of an a-C model constructed by TBMD simulation. Numbers following letters C refer to the coordination numbers of carbon atoms. (Taken with permission from Kohary and Kugler (2001). *Phys. Rev. B*, **63,** 193404. Copyright 2013 by the American Physical Society.)

thick layers over the [111] surface, the sp^3 content is high due to the memory effect. In the rest of the bulk, the sp^2 content is dominant. The partial distribution functions of a model ($T = 100$ K, $E = 1$ eV) are shown in Figure 3.19, where i and j denote the coordination numbers of the connected atoms. The average first-neighbor distances are 0.143 nm, 0.152 nm, and 0.159 nm for C3–C3, C3–C4, and C4–C4, respectively. The average coordination numbers for different models were between 3.1 and 3.4. Coordination numbers larger than four were not observed.

Angles are analyzed in detail according to the coordination numbers of their neighbors. Atoms with sp^3 local arrangements have nearly the same average values in the C3–C4–C3, C3–C4–C4, and C4–C4–C4 cases, which are around the tetrahedral angle. Clear differences are observed for sp^2 configurations. The C4–C3–C4 angles have much lower average values than the others. The latter average bond angles are close to 120°. This means that threefold-coordinated central atoms with at least two C3 first neighbors are always near planar structures, even if the third one is C4. The ring statistics for our models were also calculated. In the graphitic and diamond networks, sixfold and eightfold rings are present. The size of a ring is denoted by the number of atoms in a closed path. In our models, even-membered as well as odd-membered rings can be found, i.e. 4–7-membered rings.

Amorphous silicon model construction by molecular-dynamics simulation

Amorphous silicon networks were constructed by atomic deposition (and by rapid quenching) (Kohary and Kugler, 2004). The same molecular-dynamics computer code (ATOMDEP) as was used for the simulation of atomic bombardment was run. In the simulations, the transferable TB Hamiltonian of Kwon *et al.* (1994) was applied to describe the interaction between the silicon atoms. This group had already developed an excellent TB potential for carbon systems. All the parameters and functions of the interatomic potential for silicon were fitted to the results of the local density-functional calculations. The TB model reproduces the energies of different cluster structures, the elastic constants, the formation energies of vacancies, and interstitials in crystalline silicon. According to the authors, the only disadvantage of this TB model is that the bond lengths inside small clusters are slightly longer than those derived from other theoretical calculations or from experiments.

A surprising result was found in the ring statistics. The networks prepared by the models have a significant number of squares. Furthermore, triangles are also present in the atomic arrangements, as shown in Figure 3.20. Most of the theoretical models for a-Si do not contain such structural fractions. This seems to be an important result. It should be mentioned that a neutron-diffraction measurement carried out on a pure evaporated a-Si sample and evaluated by the RMC method led to a similar conclusion (Kugler *et al.*, 1993a). The simulation was repeated using another TB model (Lenovsky *et al.*, 1997), and similar conclusions were obtained. The a-Si structure seems to be well understood, but still newer and newer models appear each year (Gibson, 2012; Treacy and Borisenko, 2012).

Amorphous selenium model construction by molecular-dynamics simulation

Two different basic preparation methods, liquid quenching and evaporation, for a-Se are available experimentally in the laboratory. The MD computer code ATOMDEP was used to model the structure of a-Se. Two different atomic interactions were used. First, the classical empirical three-body potential of Oligschleger *et al.* (1996) was included in the computer code. A crystalline lattice cell containing 324 selenium atoms was employed to mimic the substrate.

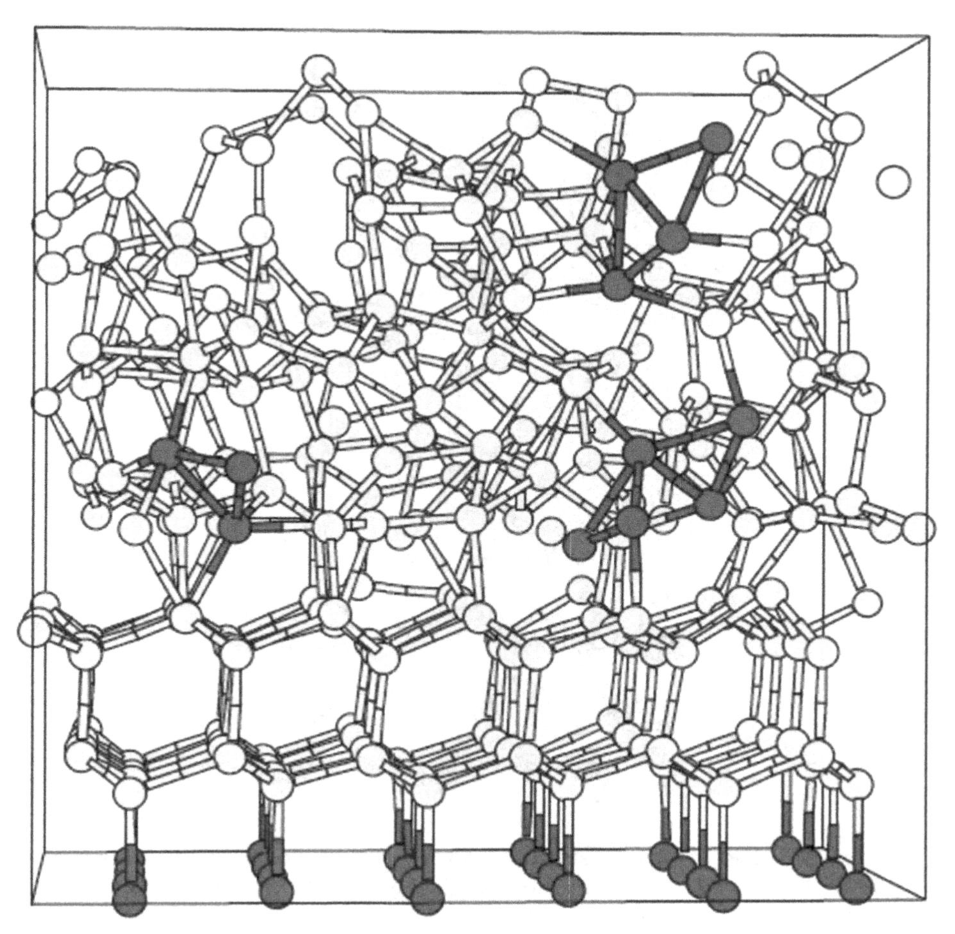

Figure 3.20. An a-Si model constructed by computer simulation of atomic deposition using the TBMD method. The atomic arrangement contains triangles (shaded).

There were 108 fixed atoms at the bottom of the substrate. The remaining atoms could move with full dynamics. The simulation cell was open along the positive z direction, and periodic boundary conditions were applied in the x and y directions. The kinetic energy of the atoms in the substrate was rescaled at every MD step ($dt = 1$ fs) to keep the substrate at a constant temperature of 100 K. In the deposition process the frequency of the atomic injection was 300 fs^{-1} (Figure 3.21).

Three different structures were constructed, with average bombarding energies of 0.1 eV, 1 eV, and 10 eV. At the end of the deposition, atomic networks contained about 1000 selenium atoms. Rapid cooling is frequently applied to construct glassy structures. The system is usually cooled down to room

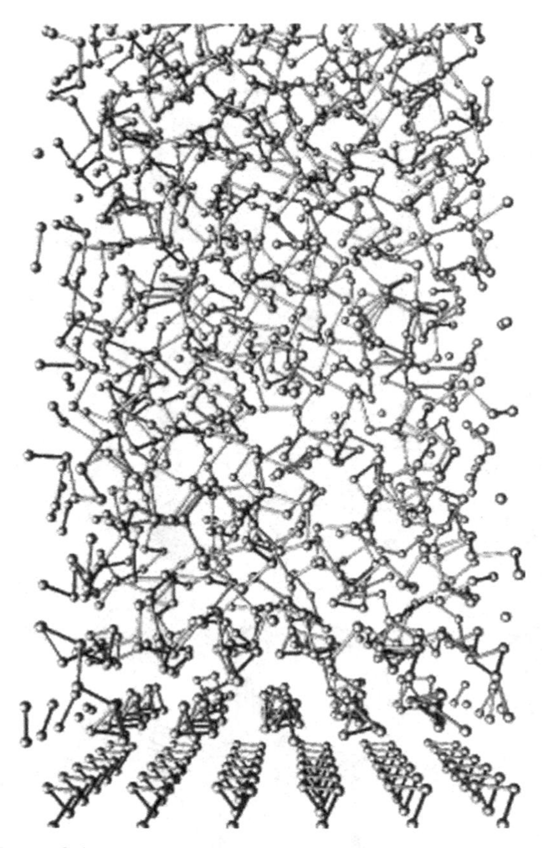

Figure 3.21. Part of the a-Se structure grown with 1 eV bombarding energy at 100 K. There are 324 substrate atoms at the bottom, of which 108 were held fixed to mimic the bulk crystal. (From Hegedus, Kohary, and Kugler (2004a). *J. Non-Cryst. Solids*, **338–340**, 283. Copyright 2013 with permission from Elsevier.)

temperature at a rate of 10^{11}–10^{16} K s^{-1} in computer simulations, although this rate is some orders of magnitude smaller in the experimental techniques. In order to retrieve information on the rapid cooling, we prepared a model in the following way. The temperature of one of the deposited films (1 eV) was increased to 900 K as an initial state (liquid phase). After this melting, the trajectories of the selenium atoms were followed by full dynamics for 100 ps. The substrate temperature maintained at 100 K leads to the cooling of the film above the substrate. This technique can be considered as the computer simulation of real splat cooling, where small droplets of melt are brought into contact with the chill-block, prepared by rapid quenching. This is more homogeneous

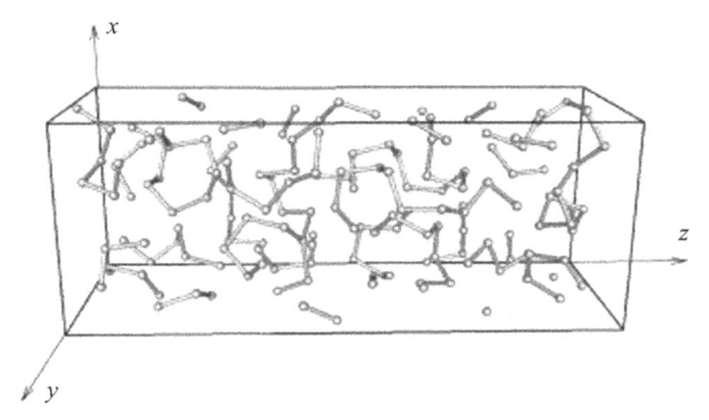

Figure 3.22. One of the several selenium glassy networks constructed using the rapid cooling technique. It contains 162 Se atoms with periodic boundary condition in the x and y directions. (From Hegedus *et al.* (2004b). *J. Non-Cryst. Solids*, **338–340**, 557. Copyright 2013 with permission from Elsevier.)

than the deposited counterpart, and is an observable difference between the two preparation techniques.

Secondly, for the description of the interaction between the selenium atoms, we used the TB model developed by Molina and co-workers (Molina *et al.*, 1999). Self-consistency was taken into account via the usual on-site Hubbard term, and was found to reduce any large charge transfer (Lomba, Molina, and Alvarez, 2000). The time step was equal to 2 fs. The temperature was controlled via the velocity-rescaling method. Several different glassy selenium networks were prepared in a rectangular box with periodic boundary conditions; samples contained 162 atoms (Figure 3.22). The "cook and quench" sample preparation is described as follows.

First, the temperature of the system was kept at 5000 K for the first 300 MD steps to randomize the atomic positions. During the next 2200 MD steps, the temperature was decreased linearly from 700 K to 250 K, driving the sample through the glass transition and reaching the condensed phase. Then the final temperature of 20 K was reached and the sample for 500 MD steps (1 ps) was relaxed. The periodic boundary conditions were lifted along the z direction at this point. This procedure provided a slab geometry containing periodic boundary conditions in only two dimensions. The system was then relaxed for another 40 000 MD steps (80 ps) at 20 K. Short quenching times in the simulation compared to those in the experiments might lead to many liquid-state defects being retained in the amorphous structure. Therefore, the Hubbard

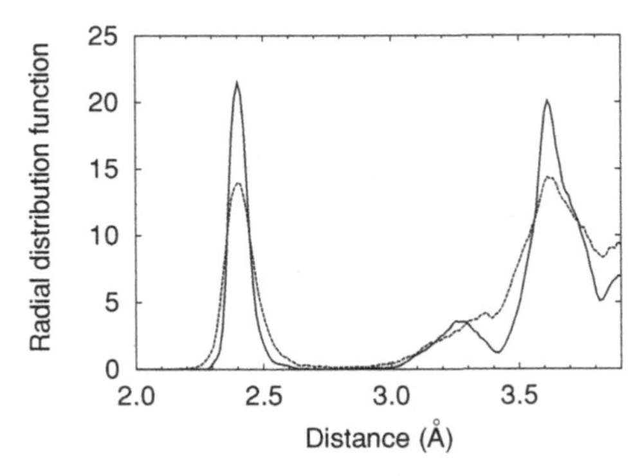

Figure 3.23. Radial distribution function of selenium glassy networks at $T =$ 20 K (solid line) and $T = 300$ K (dashed line). (Taken with permission from Hegedus *et al.* (2005). *Phys. Rev. Lett.*, **95,** 206803. Copyright 2013 by the American Physical Society.)

parameter U was taken to be 5 eV for the first 4000 MD steps during quenching to avoid a large number of coordination defects, especially onefold- and threefold-coordinated atoms. Then U was changed to its accepted value of 0.875 eV for selenium. The RDF for one of the selenium glassy networks at 20 K is shown in Figure 3.23. The first main peak at 0.24 nm belongs to covalently bonded atoms. The crystalline nearest-neighbor distance is 0.237 nm. The second peak at 0.36 nm corresponds to the intrachain second-nearest-neighbor distance as expected. The prepeak at 0.33 nm reflects the smallest interchain atomic distances in a-Se. In simulations at higher temperatures ($T = 300$ K), these two peaks merge, as observed in Figure 3.23.

Although simulations based on empirical potentials or TB models have achieved remarkable success, extension of the simulation methods to more accurate quantum mechanics would be desirable. Unfortunately, DFT within the most popular local-density approximation (LDA) is very computer time consuming – it is difficult to perform a separate self-consistent electronic minimalization at every MD step. An approach known as the Car–Parrinello method (Car and Parrinello, 1985) overcomes this difficulty, and it has been successfully applied to MD simulations of covalently bonded amorphous semiconductors, yielding valuable structural information for small systems. Structural simulations were carried out on the basis of the Car–Parinello method for a-C (Galli, Martin, Car, and Parrinello, 1989), a-C:H (Iarloni, Galli, and Martini, 1994), ta-C (Marks *et al.*, 1996), a-Si (Stich *et al.*, 1991), and a-Si:H having far fewer than 100 atoms (Buda, Chiarotti, Car, and Parrinello, 1991).

3.4 Phase change and its applications

The optical phase-change disk, i.e. the digital versatile disk (DVD), is the most successful commercialized product that uses amorphous chalcogenides (a-Chs) (see Raoux and Wuttig, 2009). It should also be emphasized that the applications of phase change are fully related to the *structural changes* of materials. The representative phase-change material is $Ge_2Sb_2Te_5$ (stoichiometric composition of $GeTe–Sb_2Te_3$), conventionally called GST. This possesses a pronounced difference in its *optical* and *electronic* properties depending upon the atomic arrangement, i.e. whether it is in the amorphous or crystalline phase. It is therefore very useful in rewritable optical storage and random access memory devices, since transitions between two states are performed over a nanosecond time range.

In rewritable optical data storage, a short pulse of a focused high-intensity laser beam locally heats GST in the crystalline state above its melting temperature (~600 °C). Rapid cooling quenches the melting spots into the amorphous phase. This "writing" process takes ~10 ns, and "reading" (detecting this spot) is achieved by reflection change using a weak-intensity laser beam. The optical reflectivity is decreased by more than 5% after reaching the amorphous state. To erase the memory spots, a laser pulse with medium power for local heating above the crystallization temperature (~150 °C) is applied. Thus the memory spots return to the original crystalline states in ~50 ns.

In electronic memory application (phase-change random access memory, PRAM), a short pulse with high-intensity voltage (or current) – instead of a laser pulse (optical recording) – is used to change the phase of materials. The local heating due to this process converts the materials into the amorphous state. "Reading" is of course performed electrically by a low-voltage pulse. The obtained amorphous state has a high resistivity, which is four orders of magnitude larger than that of the crystalline case. An intermediate moderate intensity pulse is used to heat the material above its crystallization temperature and the memory state returns to the original low-resistivity crystalline state. Note therefore that the process is similar to that of optical recording, and "heating" by *pulse* (optical or electric) is a major requirement in optical (DVD) and electronic (PRAM) processes (Atwood, 2008; Terao, Morikawa, and Ohta, 2009).

We now present an overview of the structural properties during phase changing in GST films. As the crystallization process limits the rate of the overall writing and erasing times (i.e. the rate of crystallization is slower than that for the reverse – crystalline to amorphous states – action), it is important to understand the crystallization dynamics. It should be noted, however, that the crystallization process by pulsed heating in devices may not be the same as that by isothermal heating.

Crystallization kinetics

The dynamics of crystallization in glasses has been fairly well interpreted in terms of the classical nucleation and growth model developed by Kolmogorov, Johnson and Mehl, and Avrami (the so-called KJMA model) (see, for example, Pineda and Crespo, 1999; Tonchev and Kasap, 2006). The KJMA model yields the macroscopic evolution of the transformed phase under an isothermal annealing condition with its simplicity, and it has been extended to analyze transformations under a constant heating rate. The KJMA formalism may still present some difficulties, for example fitting the KJMA results to experimental data produces unreasonable physical parameters for the crystallization, which have not been well understood so far (Senkader and Wright, 2004).

We briefly discuss now what kind of difficult problems exist in the KJMA formalism. We consider only the *macroscopic* properties, without addressing any specific *microscopic* view, which is still a matter of debate. These topics will be described later on. The KJMA theory basically assumes that the nucleation sites for the new phase are randomly distributed over the volume V, and that the growth ceases in a region where two neighboring new phase grains impinge on each other. The transformed volume V^{tr} is given by

$$dV^{tr} = \left[1 - \left(\frac{V^{tr}}{V}\right)\right] dV^{ex}. \tag{3.44}$$

Here, V^{ex} is called the extended volume, which is the *virtual* volume of the particles growing without any impingement. Integration of the above equation produces the familiar KJMA equation:

$$f = 1 - \exp\left(-f^{ex}\right), \tag{3.45}$$

where $f = V^{tr}/V$ and $f^{ex} = V^{ex}/V$, and f is the transformed volume fraction. In the case of instantaneous nucleation of spherical grains with number density N, the volume growth rate with radius r for an interface-controlled reaction can be expressed as

$$\frac{dV^{ex}}{dt} = 4\pi r^2 N V \frac{dr}{dt}. \tag{3.46}$$

Here the effective growth rate is given by $G = dr/dt$, and then f^{ex} at time t is expressed by

$$f^{ex} = \frac{4\pi}{3} N r^3 = \frac{4\pi}{3} N (Gt)^3. \tag{3.47}$$

We then obtain the well-known KJMA equation:

$$1 - f = \exp\left[-(kt)^n\right], \tag{3.48}$$

with $n = 3$. Here, n is called the Avrami exponent, and k is an effective rate constant (growth rate). The Avrami exponent should ideally be an integer providing information on the dimensionality of the crystallization process: $n = 1$ for rod-like crystals; $n = 2$ for plate-like crystals; and $n = 3$ for spherical forms.

Many experiments, however, yield numerical values that deviate significantly from the KJMA prediction. The values of n are, for example, between 1 and 5.8 in GST, and the frequency factor ν, appearing in the reaction-rate $k(T)$, takes unreasonably large values (10^{17}–10^{24} s^{-1}) (Senkader and Wright, 2004):

$$k(T) = \nu \exp\left(-\frac{E_A}{k_B T}\right), \tag{3.49}$$

where E_A is the activation energy for the phase change and T is the temperature.

Two problems that result from the KJMA formalism stated above can be overcome without losing its simplicity and beauty. The first problem is the Avrami exponent n. The theory predicts that, under heterogeneous nucleation, n represents the Euclidean dimensions ($n = 1, 2, 3$). However, the values of n are reported to be non-integer, and sometimes n is bigger than three. Non-integer values of n appear as a violation of the fundamental assumptions of the KJMA formalism, which requires: (i) a random distribution of the potential sites of nucleation, (ii) instantaneous nucleation, (iii) interface-controlled grain growth, and (iv) time- (or grain-size-) independent growth rate k.

In the following, we introduce a fractal growth concept to the space dimension (Shimakawa, 2012). Actually, $n = 2.15$ is found in supramolecular materials, and this has been attributed to the fractal growth of network aggregations (Liu and Sawant, 2002). It is generally expected that the surfaces of most aggregated materials should have a fractal dimension. The extended volume growth rate, eqn. (3.46), can then be modified as follows:

$$\frac{dV^{ex}}{dt} = C r^D N V \frac{dr}{dt}, \tag{3.50}$$

where C is a constant and D is the fractal dimension of the *surface* of the grains. Finally, we obtain f:

$$1 - f = \exp\left[-(kt)^{D+1}\right]. \tag{3.51}$$

It is therefore suggested that various values of the experimentally observed Avrami exponent n can be attributed to the complex shape of the crystalline grain surfaces with a fractal structure. This fractal-surface-controlled (FSC) crystallization may be supported by reports, for example, that $D \sim 3.0$ for silica gels and $D = 2.6$–2.9 for sandstones (Shimakawa, 2012).

The second problem is the temperature-dependent rate constant $k(T)$ (see eqn. (3.49)). The reported activation energies E_A, for example in GST, are around 2 eV (1−3 eV). We recall that unreasonably large frequency factors ($10^{17}-10^{24}$ s^{-1}), corresponding to 10^2–10^9 eV, can be deduced from the application of the classical KJMA model. These ν values are clearly unacceptable. What is the origin of these unreasonable values for the frequency factor? Many thermally activated phenomena, such as electronic and ionic transport, and structural relaxation in disordered matter, obey the Meyer–Neldel compensation rule (Yelon, Movaghar, and Crandall, 2006). In eqn. (3.49), $k(T)$ should be modified as follows:

$$k(T) = \nu_0 \exp\left(\frac{E_A}{E_{MN}}\right) \exp\left(-\frac{E_A}{k_B T}\right), \tag{3.52}$$

where ν is given by

$$\nu \equiv \nu_0 \exp\left(\frac{E_A}{E_{MN}}\right), \tag{3.53}$$

where E_{MN} is the Meyer–Neldel energy, and ν_0 is a constant. Equation (3.52) is called the Meyer–Neldel (MN) rule or compensation law (Yelon *et al.*, 2006).

Introduction of the fractal geometry and the MN rule on thermally activated processes (extended KJMA) overcomes some of the difficulties with the classical KJMA model, although the understanding of crystallization from the microscopic point of view is still lacking. In particular, for GST, due to its extraordinary fast crystallization rate and small area, the microscopic structural changes during nucleation and growth are not easy to detect experimentally.

Ab initio molecular-dynamics (AIMD) simulations based on density-functional theory have been applied to see what happens in the early stage of crystallization of GST (Lee and Elliott, 2011). The simulations confirm the stochastic nature of the onset of crystallization as assumed in classical nucleation theory. An incubation period for crystallization is clearly defined on annealing at 600 K. The critical crystal molecules are estimated to comprise five to ten chemically ordered $(Ge,Sb)_4Te_4$ units. Tellurium atoms in the crystalline phase form a perfect fcc sublattice of rocksalt structure, and Ge or Sb atoms occupy sites in the other sublattice. The simulations make clear that the formation of ordered planar structures in the amorphous phase plays a critical role in lowering the interfacial energy between crystalline clusters and the amorphous phase, yielding ultrafast crystallization in GST. The above prediction from the AIMD simulations is very important, and therefore the simulations will be discussed in the following.

Structural transformations resulting from phase changes in GST

In the previous subsection, we discussed the early stage of the crystallization process via AIMD simulations (Lee and Elliott, 2011). Now we discuss the structure of GST, in particular the structural difference between its amorphous and crystalline states. We need to understand why GST shows fast phase transitions over a nanosecond time scale. Many structural models have been proposed to interpret the fast phase transitions (Kolobov *et al.*, 2004; Hegedus and Elliott, 2008; Elliott (2009); Yamada, 2009). Although the detailed nature of GST structure is still disputed, the recently suggested resonance bonding model may be useful in understanding the difference between the crystalline and amorphous states in GST (Shportko *et al.*, 2008). It should be noted, however, that there is no clear evidence of resonance bonding.

We briefly review the principal structural models. The common understanding of the GST structure is that crystalline GST has a (distorted) rocksalt structure and that the Ge environment is fourfold coordinated in the amorphous phase. There is still disagreement on the detailed structure among the modeling studies. Kolobov *et al.* (2004) have proposed the so-called *umbrella flip* model from the x-ray absorption near-edge structure (XANES) spectra. While an fcc-like structure is retained in both the crystalline and re-amorphized states, the Ge atom occupies octahedral and tetrahedral symmetry positions in the re-crystallized and re-amorphized states. An intense laser pulse causes the longer bonds to break, and the Ge atom moves into the tetrahedral symmetry position. A molten (liquid) state as a transit path is not required from the crystalline to the amorphous states in this model.

Yamada (2009) takes another view in the subject of amorphization. The crystalline phases (distorted rocksalt structure) show very large thermal vibrations at high temperatures. This increases the randomness of the atomic distributions, which is rather similar to that of the liquid phase. The amorphous structure reflects the very isotropic atomic configuration in the liquid state. Thus the atomic arrangements for both crystalline and amorphous states are very close, and the transit time for phase change should be very short.

Finally, we discuss the very large static dielectric constant $\varepsilon(0)$ in the crystalline state of GST. It is twice as large for the crystalline phase as it is for the amorphous phase. This difference is attributed to a significant change in bonding between the two phases (Shportko *et al.*, 2008). The optical dielectric constant of the amorphous phase is that expected of a covalent bonding (following the $8 - N$ rule), whereas that of the crystalline phase is expected to be strongly enhanced by *resonant bonding* effects. The existence of resonance bonding requires a longer range order than the conventional covalent bond length, whereas an electron pair requires only the ordering of the nearest neighbors. Note that the second- and higher-order neighbors should be aligned in resonance bonding. As a higher $\varepsilon(0)$ is also expected for smaller bandgap materials (Phillips, 1973), the existence of resonant bonds may not be the condition necessary to produce a higher dielectric constant.

A unique feature of the crystalline state of phase-change materials (not only GST) is based on a distorted cubic structure. Various novel techniques discussed in this chapter may clarify this complex issue. A recent monograph (Kolobov and Tominaga, 2012) can provide a deeper understanding of phase-change materials for readers.

References

Abell, G.C. (1985). Empirical chemical pseudopotential theory of molecular and metallic bonding. *Phys. Rev. B*, **31**, 6184–6196.

Adam, G. and Gibbs, J.H. (1965). On the temperature dependence of cooperative relaxation properties in glass-forming liquids. *J. Phys. Chem.*, **43**, 139–146.

Allen, F.H., Kennard O., and Taylor, R. (1983). Systematic analysis of structural data as a research technique in organic chemistry. *Acc. Chem. Res.*, **16**, 146–153.

Anderson, P.W., Halperin, B.I., and Varma, C.M. (1972). Anomalous low-temperature thermal properties of glasses and spin glasses. *Phil. Mag.*, **25**, 1–9.

Angell, C.A. (1988). Structural instability and relaxation in liquid and glassy phases near the fragile liquid limit. *J. Non-Cryst. Solids*, **102**, 205–221.

Angell, C.A. (1995). Formation of glasses from liquids and biopolymers. *Science*, **267**, 1924–1935.

Aniya, M. (2002). A model for the fragility of the melts. *J. Therm. Anal. Calorim.*, **69**, 971–978.

Atwood, G. (2008). Phase-change materials for electronic memories. *Science*, **321**, 210–211.

Barna, A., Barna, P.B., Radnoczi, G., Toth, L., and Thomas, P. (1977). A comparative study of the structure of evaporated and glow discharge silicon. *Phys. Stat. Sol. A*, **41**, 81–84.

Beeman, D. (1976). Some multistep methods for use in molecular dynamics calculations. *J. Comput. Phys.*, **20**, 130–139.

Bellissent, R., Menelle, A., Howells, W.S. *et al.* (1989). The structure of amorphous Si:H using steady state and pulsed neutron sources. *Physica B*, **156–157**, 217–219.

Bose, S.K., Winer, K., and Andersen, O.K. (1988). Electronic properties of a realistic model of amorphous silicon. *Phys. Rev. B*, **37**, 6262–6277.

Brey, L., Tejedor, C., and Verges, J.A. (1984). Comment on "Static charge fluctuations in amorphous silicon". *Phys. Rev. Lett.*, **52**, 1840.

Buda, F., Chiarotti, G.L., Car, R., and Parrinello, M. (1991). Structure of hydrogenated amorphous silicon from *ab initio* molecular dynamics. *Phys. Rev.*, **B44**, 5908–5911.

Car, R. and Parrinello, M. (1985). Unified approach for molecular dynamics and density-functional theory. *Phys. Rev. Lett.*, **55**, 2471–2474.

Chen, W-H., Wu, C-H., and Cheng, H-C. (2011). Modified Nosé–Hoover thermostat for solid state for constant temperature molecular dynamics simulation. *J. Comput. Phys.*, **230**, 6354–6366.

Cooper, N.C., Goringe C.M., and McKenzie, D.R. (2000). Density functional theory modelling of amorphous silicon. *Comput. Mater. Sci.*, **17**, 1–6.

Cser, L., Török, Gy., Krexner, G., Sharkov, I., and Faragó, B. (2002). Holographic imaging of atoms using thermal neutrons. *Phys. Rev. Lett.*, **89**, 175504, 1–4.

Cuello, G.J. (2008). Structure factor determination of amorphous materials by neutron diffraction. *J. Phys.: Condens. Matter*, **20**, 244109, 1–9.

Ding, K. and Andersen, H.C. (1986). Molecular-dynamics simulation of amorphous germanium. *Phys. Rev. B*, **34**, 6987–6991.

Drabold, D.A., Fedders, P.A., Klemm, S., and Sankey, O.F. (1991). Finite-temperature properties of amorphous silicon. *Phys. Rev. Lett.*, **67**, 2179–2182.

Elliott, S.R. (1990). *Physics of Amorphous Materials*, 2nd edn. Harlow: Longman Scientific & Technical.

Elliott, S.R. (2009). Structure of amorphous Ge-Sb-Te solids. In *Phase Change Materials: Science and Applications*, eds. S. Raoux and M. Wuttig. New York: Springer, chap. 4.

Etherington, J.H., Wright, A.C., Wenzel, J.T., Dore, J.C., Clarke, J.H., and Sinclair, R.N. (1982). A neutron diffraction study of the structure of evaporated amorphous germanium. *Non-Cryst. Solids*, **48**, 265–289.

Feynman, R.P. (1939). Forces in molecules. *Phys. Rev.*, **56**, 340–343.

Fortner, J. and Lannin, J.S. (1989). Radial distribution functions of amorphous silicon. *Phys. Rev. B*, **39**, 5527–5530.

Freitas, R.J., Shimakawa, K., and Kugler, S. (2013). Some remarks on glass-transition temperature in chalcogenide glasses: a correlation with the microhardness. *Chalcogenide Lett.*, **10**, 39–43.

Galli, G., Martin, R.M., Car, M., and Parrinello, M. (1989). Structural and electronic properties of amorphous carbon. *Phys. Rev. Lett.*, **62**, 555–558.

Gaskell, P.H., Saeed, A., Chieux, P., and McKenzie, D.R. (1991). Neutron-scattering studies of the structure of highly tetrahedral amorphous diamondlike carbon. *Phys. Rev. Lett.*, **67**, 1286–1289.

Gereben, O. and Pusztai, L. (1994). Structure of amorphous semiconductors: reverse Monte Carlo studies on a-C, a-Si, and a-Ge. *Phys. Rev. B*, **50**, 14136–14143.

Gereben, O. and Pusztai, L. (2012). RMC_POT, a computer code for reverse Monte Carlo modeling the structure of disordered systems containing molecules of arbitrary complexity. *J. Comput. Chem.*, **33**, 2285–2291.

Gereben, O., Jóvári, P., Temleitner, L., and Pusztai, L. (2007). A new version of the RMC++ Reverse Monte Carlo programme, aimed at investigating the structure of covalent glasses. *J. Optoelectron. Adv. Mater.*, **9**, 3021–3027.

Gibson, J.M. (2012). Solving amorphous atructures – two pairs beat one. *Science*, **335**, 929–930.

Gilkes, K.W.R., Gaskell, P.H., and Robertson, J. (1995). Comparison of neutron-scattering data for tetrahedral amorphous carbon with structural models. *Phys. Rev. B*, **51**, 12303–12312.

Goodwin, L., Skinner, A.J., and Pettifor, D.G. (1989). Generating transferable tight-binding parameters: application to silicon. *Europhys. Lett.*, **9**, 701–706.

Greaves, G.N. and Sen, S. (2007). Inorganic glasses, glass-forming liquids and amorphizing solids. *Adv. Phys.*, **56**, 1–166.

Guttman, L., Ching, W.Y., and Rath, J. (1980). Charge-density variation in a model of amorphous silicon. *Phys. Rev. Lett.*, **44**, 1513–1516.

Hegedus, J. and Elliott, S.R. (2008). Microscopic origin of the fast crystallization ability of Ge-Sb-Te phase-change memory materials. *Nature Mater.*, **7**, 399–405.

Hegedus, J. and Kugler, S. (2005). Growth of amorphous selenium thin films: classical versus quantum mechanical molecular dynamics simulation. *J. Phys.: Condens. Matter*, **17**, 6459–6468.

Hegedus, J., Kohary, K., and Kugler, S. (2004a). Comparative analysis of different preparation methods of chalcogenide glasses: molecular dynamics structure simulations. *J. Non-Cryst. Solids*, **338–340**, 283–286.

Hegedus, J., Kohary, K., Kugler, S., and Shimakawa, K. (2004b). Photo-induced volume changes in selenium. Tight-binding molecular dynamics study. *J. Non-Cryst. Solids*, **338–340**, 557–560.

Hegedus, J., Kohary, K., Pettifor, D.G., Shimakawa, K., and Kugler, S. (2005). Photo-induced volume changes in amorphous selenium. *Phys. Rev. Lett.*, **95**, 206803, 1–4.

Hellmann, H.G.A. (1933). Zur rolle der kinetischen Elektronenenergie für die zweischenatomaren Krafte. *Z. Phys.*, **85**, 180–190.

Hensel, H., Klein, P., Urbassek, H.M., and Frauenheim, T. (1996). Comparison of classical and tight-binding molecular dynamics for silicon growth. *Phys. Rev. B*, **53**, 16497–16503.

Herrero, C.P. (2000). Quantum atomic dynamics in amorphous silicon; a path-integral Monte-Carlo simulation. *J. Phys.: Condens. Matter*, **12**, 265–274.

Hoover, W.G. (1985). Canonical dynamics: equilibrium phase-space distributions. *Phys. Rev. A*, **31**, 1695–1697.

Iarloni, S., Galli, G., and Martini, O. (1994). Microscopic structure of hydrogenated amorphous carbon. *Phys. Rev. B*, **49**, 7060–7063.

Ishimaru, M. (2001). Molecular-dynamics study on atomistic structures of amorphous silicon. *J. Phys.: Condens. Matter*, **13**, 4181–4189.

Ishimaru, M., Munetoh, S., and Motooka, T. (1997). Generation of amorphous silicon structures by rapid quenching: a molecular-dynamics study. *Phys. Rev. B*, **56**, 15133–15138.

Ishimaru, M., Yamaguchi, M., and Hirotsu, Y. (2004). Structural relaxation of amorphous silicon-germanium alloys: molecular-dynamics study. *Jpn. J. Appl. Phys.*, **43**, 7966–7970.

Jian, Z., Kaiming, Z., and Xide, X. (1990). Modification of Stillinger-Weber potentials for Si and Ge. *Phys. Rev. B*, **41**, 12915–12918.

Jóvári, P., Delaplane, R.G., and Pusztai, L. (2003). Structural models of amorphous selenium. *Phys. Rev. B*, **67**, 172201, 1–4.

Justo, J.F., Bazant, M.Z., Kaxiras, E., Bulatov, V.V., and Yip, S. (1998). Interatomic potential for silicon defects and disordered phases. *Phys. Rev. B*, **58**, 2539–2550.

Kakinoki, J., Katada, K., Hanawa, T., and Ino, T. (1960). Electron diffraction study of evaporated carbon film. *Acta Cryst.*, **13**, 171–179.

Kauzmann, W. (1948). The nature of glassy state and the behaviour of liquids at low temperatures. *Chem. Rev.*, **43**, 219–256.

Keating, P.N. (1966). Effect of invariance requirements on the elastic strain energy of crystals with application to the diamond structure. *Phys. Rev.*, **145**, 637–645.

Kelires, P.C. and Tersoff, J. (1988). Glassy quasithermal distribution of local geometries and defects in quenched amorphous silicon. *Phys. Rev. Lett.*, **61**, 562–565.

Klug, D.D. and Whalley, E. (1982). Effective charges of amorphous silicon, germanium, arsenic, and ice. *Phys. Rev. B*, **25**, 5543–5546.

Kohary, K. and Kugler, S. (2001). Growth of amorphous carbon. Low energy molecular dynamics simulation of atomic bombardment. *Phys. Rev. B*, **63**, 193404, 1–4.

Kohary, K. and Kugler, S. (2004). Growth of amorphous silicon: low energy molecular dynamics simulation of atomic bombardment. *Mol. Simul.*, **30**, 17–22.

Kolobov, A and Tominaga, J. (2012). *Chalcogenides: Metastability and Phase Change Phenomena*. Berlin and Heidelberg: Springer.

Kolobov, A.V, Fons, P., Frenkel, A.I., and Ankudinov, A.L. (2004). Understanding of the phase-change mechanism of rewritable optical media. *Nature Mat.*, **3**, 703–708.

Kramer, B., King, H., and Mackinnon, A. (1983). Charge fluctuations in hydrogenated amorphous silicon. *J. Non-Cryst. Solids*, **59–60**, 73–76.

Kugler, S. and Náray-Szabó, G. (1987). Charge distribution in amorphous silicon clusters: quantum chemical study combined with ring statistics. *J. Non-Cryst. Solids*, **97–98**, 503–506.

Kugler, S. and Náray-Szabó, G. (1991a). Weak bonds and atomic charge distribution in hydrogenated amorphous silicon. *J. Non-Cryst. Solids*, **137–138**, 295–298.

Kugler, S. and Náray-Szabó, G. (1991b). Atomic charge distribution in diamondlike amorphous carbon. *Jpn. J. Appl. Phys.*, **30**, L1149–L1151.

Kugler, S. and Varallyay, Z. (2001). Possible unusual atomic arrangements in the structure of amorphous silicon. *Phil. Mag. Lett.*, **81**, 569–574.

Kugler, S., Surjan, P.R., and Náray-Szabó, G. (1988). Theoretical estimation of static charge fluctuation in amorphous silicon. *Phys. Rev. B*, **37**, 9069–9071.

Kugler, S., Molnar, G., Peto, G. *et al.* (1989). Neutron-diffraction study of the structure of evaporated pure amorphous silicon. *Phys. Rev. B*, **40**, 8030–8032.

Kugler, S., Pusztai, L., Rosta, L., Chieux, P., and Bellissent, R. (1993a). Structure of evaporated pure amorphous silicon: neutron-diffraction and reverse Monte Carlo investigation. *Phys. Rev. B*, **48**, 7685–7688.

Kugler, S., Shimakawa, K., Watanabe, T., Hayashi, K., Laszlo, I., and Bellissent, R. (1993b). The temperature dependence of amorphous carbon structure. *J. Non-Cryst. Solids*, **164–166**, 1143–1146.

Kwon, I., Biswas, R., Wang, C.Z., Ho, K.M., and Soukoulis, C.M. (1994). Transferable tight-binding models for silicon. *Phys. Rev. B*, **49**, 7242–7250.

Laaziri, K., Kycia, S., Roorola, S. *et al.* (1999). High resolution radial distribution function of pure amorphous silicon. *Phys. Rev. Lett.*, **82**, 3460–3463.

Langer, J.S. (2006). Excitation chains at the glass transition. *Phys. Rev. Lett.*, **97**, 11504, 1–4.

Laszlo, I. (1998). Formation of cage-like C_{60} clusters in molecular dynamics simulations. *Europhys. Lett.*, **44**, 741–746.

Laszlo, I. (1999). Molecular dynamics study of the C_{60} molecule. *J. Mol. Struct.: THEOCHEM*, **463**, 181–184.

Lee, T.H. and Elliott, S.R. (2011). *Ab initio* computer simulation of the early stage of the crystallization: application to $Ge_2Sb_2Te_5$ phase-change materials. *Phys. Rev. Lett.*, **107**, 145702, 1–5.

Lenovsky, T., Kress, J.D., Kwon, I. *et al.* (1997). Highly optimized tight-binding model of silicon. *Phys. Rev. B*, **55**, 1528–1544.

Ley, L., Reichardt, J., and Johnson, R.L. (1982). Static charge fluctuations in amorphous silicon. *Phys. Rev. Lett.*, **49**, 1664–1667.

Li, F. and Lannin, J.S. (1990). Radial distribution function of amorphous carbon. *Phys. Rev. Lett.*, **65**, 1905–1908.

Liu, X.Y. and Sawant, P.D. (2002). Mechanism of the formation of self-organized microstructures in soft functional materials. *Adv. Mater.*, **14**, 421–426.

Lomba, E., Molina, D., and Alvarez, M. (2000). Hubbard corrections in a tight-binding Hamiltonian for Se: effects on the band structure, local order, and dynamics. *Phys. Rev. B*, **61**, 9314–9321.

Lukács, R. and Kugler, S. (2010). A simple model for the estimation of charge accumulation in amorphous selenium. *Chem. Phys. Lett.*, **494**, 287–288.

McGreevy, R.L. and Pusztai, L. (1988). Reverse Monte Carlo simulation: a new technique for the determination of disordered structures. *Molec. Sim.*, **1**, 359–367.

Marks, N.A., McKenzie, D.R., Pailthorpe B.A., Bernasconi, M., and Parrinello, M. (1996). Microscopic structure of tetrahedral amorphous carbon. *Phys. Rev. Lett.*, **76**, 768–771.

Metropolis, N., Rosenbluth, A.W., Rosenbluth, M.N., Teller, A.H., and Teller, E. (1953). Equation of state calculations by fast computing machines. *J. Chem. Phys.*, **21**, 1087.

Molina, D., Lomba, E., and Kahl, G. (1999). Tight-binding model of selenium disordered phases. *Phys. Rev. B*, **60**, 6372–6382.

Moss, S.C. and Graczyk, J.F. (1969). Evidence of voids within the as-deposited structure of glassy silicon. *Phys. Rev. Lett.*, **23**, 1167–1171.

Nose, S. (1984). A unified formulation of the constant temperature molecular-dynamics methods. *J. Chem. Phys.*, **81**, 511–519.

Novikov, V.N. and Sokolov, A.P. (2013). Role of quantum effects in the glass transition. *Phys. Rev. Lett.*, **110**, 065701, 1–5.

Oligschleger, C., Jones, R.O., Reimann, S.M., and Schober, H.R. (1996). Model interatomic potential for simulations in selenium. *Phys. Rev. B*, **35**, 6165–6173.

Overney, G., Zhong, W., and Tománek, D. (1993). Structural rigidity and low frequency vibrational modes of long carbon tubules. *Z. Phys. D*, **27**, 93–96.

Phillips, J.C. (1973). *Bonds and Bands in Semiconductors*. New York and London: Academic Press.

Phillips, W.A. (1972). Tunneling states and the low-temperature thermal expansion of glasses. *J. Low Temp. Phys.*, **7**, 351–360.

Pineda, E.P. and Crespo, D. (1999). Microstructure development in Kolmogorov, Johnson-Mehl, and Avrami nucleation and growth kinetics. *Phys. Rev. B*, **60**, 3104–3112.

Pizzagalli, L., Godet, J., Guénolé, J. *et al.* (2013). A new parametrization of the Stillinger–Weber potential for an improved description of defects and plasticity of silicon. *J. Phys.: Condens. Matter*, **25**, 055801, 1–12.

Powell, D., Migliorato, M.A., and Cullis, A.G. (2007). Optimized Tersoff potential parameters for tetrahedrally bonded III-V semiconductors. *Phys. Rev. B*, **75**, 115202, 1–9.

Pusztai, L. and Kugler S. (2005). Comparison of the structures of evaporated and ion-implanted amorphous silicon samples. *J. Phys.: Condens. Matter*, **17**, 2617–2624.

Raoux, S. and Wuttig, M. (2009). Preface. In *Phase Change Materials: Science and Applications*, eds. S. Raoux and M. Wuttig. New York: Springer.

Rücker, H. and Methfessel, M. (1995). Anharmonic Keating model far group-IV semiconductors with application to the lattice dynamics in alloys of Si, Ge, and C. *Phys. Rev. B*, **52**, 11059–11072.

Sears, V.F. (1992). Neutron scattering lengths and cross-sections. *Neutron News*, **3**(3), 26–37.

Senkader, S. and Wright, C.D. (2004). Models for phase-change of $Ge_2Sb_2Te_5$ in optical and electrical memory devices. *J. Appl. Phys.*, **95**, 504–511.

Shimakawa, K. (2012). Dynamics of crystallization with fractal geometry: extended KJMA approach in glasses. *Phys. Status Solidi B*, **249**, 2024–2027.

Shimakawa, K., Hayashi, K., Kameyama, T., Watanabe, T., and Morigaki K. (1991). Anomalous electrical conduction in graphite-vaporized films. *Phil. Mag. Lett.*, **64**, 375–378.

Shportko, K., Kremers, S., Woda, M., Lencer, D., Robertson, J., and Wuttig, M. (2008). Resonant bonding in crystalline phase-change materials. *Nature Mat.*, **7**, 653–658.

Sim, E., Beckers, J., Leeuw, S., Thorpe, M., and Ratner, M.A. (2005). Parameterization of an anharmonic Kirkwood-Keating potential for $Al_xGa_{(1-x)}$ As alloy. *J. Chem. Phys.*, **122**, 174702, 1–6.

Stich, I., Car, R., and Parrinello, M. (1991). Amorphous silicon studied by *ab initio* molecular dynamics: preparation, structure, and properties. *Phys. Rev. B*, **44**, 11092–11104.

Stillinger, F.H. and Weber, T.A. (1985). Computer simulation of local order in condensed phases of silicon. *Phys Rev. B*, **31**, 5262–5271.

Stone, A.J. and Wales, D.J. (1986). Theoretical studies of icosahedral C_{60} and some related species. *Chem. Phys. Lett.*, **128**, 501–503.

Swope, W.C., Andersen, H.C., Berens, P.H., and Wilson, K.R. (1982). A computer simulation method for the calculation of equilibrium constants for the formation of physical clusters of molecules: application to small water clusters. *J. Chem. Phys.*, **76**, 637–649.

Tabuchi, N., Kawahara, T., Arai, T., Morimoto, J., and Matsumura, H. (2004). Development of structural analysis method based on reverse Monte Carlo simulation and its application to catalytic chemical vapor deposition hydrogenated amorphous silicon. *Jpn. J. Appl. Phys.*, **43**, 6873–6879.

Tanaka, K. (1985). Glass transition of covalent glasses. *Solid State Commun.*, **54**, 867–869.

Tanaka, K. and Shimakawa, K. (2011). *Amorphous Chalcogenide Semiconductors and Related Materials*. New York: Springer.

Tang, M.S., Wang, C.Z., Chan, C.T., and Ho, K.M. (1996). Environment-dependent tight-binding potential model. *Phys. Rev. B*, **53**, 979–982.

Tegze, M. and Faigel, Gy. (1996). X-ray holography with atomic resolution. *Nature*, **380**, 49–51.

Terao, M., Morikawa, T., and Ohta, T. (2009). Electrical phase-change memory: fundamentals and state of the art. *Jpn. J. Appl. Phys.*, **48**, 080001, 1–14.

Tersoff, J. (1986). New empirical model for the structural properties of silicon. *Phys. Rev. Lett.*, **56**, 632–635.

Tersoff, J. (1988a). New empirical approach for the structure and energy of covalent systems. *Phys. Rev. B*, **37**, 6991–7000.

Tersoff, J. (1988b). Empirical interatomic potential for silicon with improved elastic properties. *Phys. Rev. B*, **38**, 9902–9905.

Tersoff, J. (1988c). Empirical interatomic potential for carbon, with application to amorphous carbon. *Phys. Rev. Lett.*, **61**, 2879–2882.

Tersoff, J. (1989). Modeling solid-state chemistry: interatomic potentials for multicomponent systems. *Phys. Rev. B*, **39**, 5566–5568. (Erratum: (1990) *Phys. Rev. B*, **41**, 3248.)

Tichý, L. and Tichá, H. (1995). Covalent bond approach to the glass-transition temperature of chalcogenide glasses. *J. Non-Cryst. Solids*, **189**, 141–146.

Tonchev, D. and Kasap, S.O. (2006). Thermal properties and thermal analysis: fundamentals, experimental techniques and applications. In *The Springer Handbook of Electronic and Photovoltaic Materials*, eds. S.O. Kasap and P. Capper. Heidelberg: Springer, chap. 19.

Tóth, G. and Náray-Szabó, G. (1994). Novel semiempirical method for quantum Monte Carlo simulation: application to amorphous silicon. *J. Chem. Phys.*, **100**, 3742–3746.

Treacy, M.M.J. and Borisenko, K.B. (2012). The local structure of amorphous silicon. *Science*, **335**, 950–953.

Valladares, A.A., Alvarez, F., Liu, Z., Sticht, J., and Harris, J. (2001). *Ab initio* studies of the atomic and electronic structure of pure and hydrogenated a-Si. *Eur. Phys. J. B*, **22**, 443–453.

Verlet, L. (1967). Computer "experiments" on classical fluids. I. Thermodynamical properties of Lennard–Jones molecules. *Phys. Rev.*, **159**, 98–103.

Vink, R.L.C., Barkema, G.T., van der Weg, W.F., and Mousseau, N. (2001). Fitting the Stillinger–Weber potential to amorphous silicon. *J. Non-Cryst. Solids*, **282**, 248–255.

Walters, J.K., Gilkes, K.W.R., Wicks, J.D., and Newport, R.J. (1998). Progress in modelling the chemical bonding in tetrahedral amorphous carbon. *Phys. Rev. B*, **58**, 8267–8276.

Wang, C.Z., Pan, B.C., and Ho, K.M. (1999). An environment-dependent tight-binding potential for Si. *J. Phys.: Condens. Matter*, **11**, 2043–2049.

Wilson, M. and Salmon, P.S. (2009). Network topology and the fragility of tetrahedral glass-forming liquids. *Phys. Rev. Lett.*, **103**, 157801, 1–4.

Winer, K. and Cardona, M. (1986). Theory of infrared absorption in amorphous silicon. *Solid State Commun.*, **60**, 207–211.

Wooten, F. and Weaire, D. (1984). Generation of random network models with periodic boundary conditions. *J. Non-Cryst. Solids*, **64**, 325–334.

Wooten, F., Winer, K., and Weaire, D. (1985). Computer generation of structural models of amorphous Si and Ge. *Phys. Rev. Lett.*, **54**, 1392–1395.

Xu, C.H., Wang, C.Z., Chan, C.T., and Ho, K.M. (1992). A transferable tight-binding potential for carbon. *J. Phys.: Condens. Matter*, **4**, 6047–6057.

Yamada, N. (2009). Development of materials for third generation optical strange media. In *Phase Change Materials: Science and Applications*, eds. S. Raoux and M. Wuttig. New York: Springer, chap. 10.

Yang, R. and Singh, J. (1998). Study of the stability of hydrogenated amorphous silicon using tight-binding molecular dynamics. *J. Non-Cryst. Solids*, **240**, 29–34.

Yelon, A., Movaghar, B., and R.S. Crandall, R.S. (2006). Multi-excitation entropy: its role in thermodynamics and kinetics. *Rep. Prog. Phys.*, **69**, 1145–1194.

4

Electronic structure

The influence of disorder causes essential differences in the electronic density of states (DOS) of amorphous and crystalline semiconductors. A description of the electronic properties starts with an understanding of the covalent bonds in amorphous semiconductors. Defects in these materials also modify their optical and electronic properties. Different defects and their influences are overviewed in the second part of this chapter.

4.1 Chemical bonds

In saturated covalently bonded semiconductors, the $8 - N$ rule states that the number of sigma bonds or the local coordination number equals eight minus the relevant column (group) number of the periodic table. This rule is realized within a valence bond framework by assuming that single saturated covalent bonds are formed with next neighbors, establishing the closed and stable octet shell of electrons. The only exception to this rule is graphitic carbon, which is a good conductor. For $N = 4$, germanium, silicon, and diamond-like carbon atoms take the tetrahedral diamond local structure with four sigma bonds, and for $N = 5$ the pnictides arsenic, antimony, and bismuth take a puckered layer structure with three sigma bonds. Furthermore, for $N = 6$, the chalcogenides selenium and tellurium take helical chain structures having two sigma bonds. Finally, group VII elements form diatomic molecules, i.e. atoms having only one electron for the sigma bond.

Group IV elements

The framework of the linear combination of atomic orbitals (LCAO) can be used to describe the electronic structure and bonds in these materials.

Two-center-localized molecular orbitals are constructed using the LCAO form of the sigma bonds. We illustrate this methodology for carbon. An isolated carbon atom has the electron configuration $1s^2 2s^2 2p^2$. Therefore, one could naively think the deeper core states (for example the $1s$ and $2s$ states) are not involved in bonding, and that twofold chemical bonds are formed using the $2p$ states only. However, in the majority of cases the situation is more complex, and the $2s$ states also contribute to bond formation; this mechanism is known as *hybridization*.

In covalently bonded semiconductors the valence electrons are localized and form the chemical bonds. Therefore the valence electron wave functions are similar to the bonding orbitals found in molecules. For a tetrahedrally coordinated carbon, for example methane (CH_4), the carbon atom should have four orbitals with the correct symmetry to bond to the four hydrogen atoms. The solutions, linear combinations of the $2s$ and $2p$ wave functions, are known as hybridized orbitals. Therefore, in the case of tetrahedral carbon, four orbitals are required. The $2s$ orbital is mixed with the three $2p$ orbitals to form four sp^3 hybrids:

$$\begin{aligned}
sp^3 &= \frac{1}{2}s + \frac{1}{2}p_x + \frac{1}{2}p_y + \frac{1}{2}p_z, \\
sp^3 &= \frac{1}{2}s - \frac{1}{2}p_x - \frac{1}{2}p_y + \frac{1}{2}p_z, \\
sp^3 &= \frac{1}{2}s - \frac{1}{2}p_x + \frac{1}{2}p_y - \frac{1}{2}p_z, \\
sp^3 &= \frac{1}{2}s + \frac{1}{2}p_x - \frac{1}{2}p_y - \frac{1}{2}p_z.
\end{aligned} \tag{4.1}$$

These four hybrids are orthonormalized states. In CH_4, four sp^3 hybridized orbitals are overlapped by the hydrogen $1s$ orbitals, yielding four sigma bonds (i.e. four single covalent bonds). The four equivalent covalent bonds are of the same length and strength. When the atoms from group IV of the periodic table combine to combine to form a condensed phase, the interaction splits the valence states into the electron-bonding and higher-energy antibonding levels. Between these two levels a gap can be found where no electron states exist. In a crystal, the sp^3 orbitals overlap with similar orbitals of the four adjacent atoms, and a well-defined sigma-bond network is formed, as shown in Figure 4.1. All the bonds are occupied by two electrons. This theory also fits the fourfold-coordinated group IV elements (such as diamond-like carbon, silicon, and germanium) in non-crystalline arrangements. However, the ½ prefactors in eqn. (4.1) have to be modified in order to account for deviations from the ideal tetrahedral arrangement of four nearest-neighbor atoms.

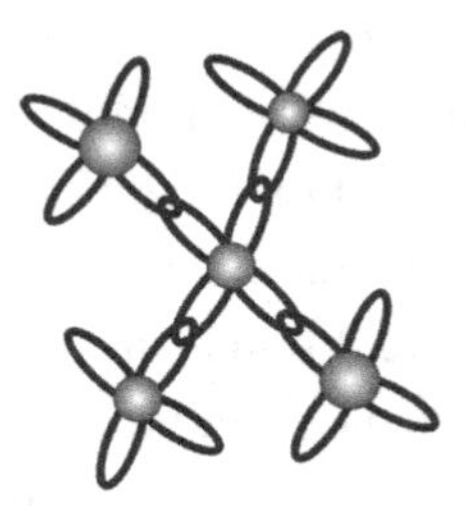

Figure 4.1. The sp^3 orbitals overlap with the similar orbitals of four adjacent atoms.

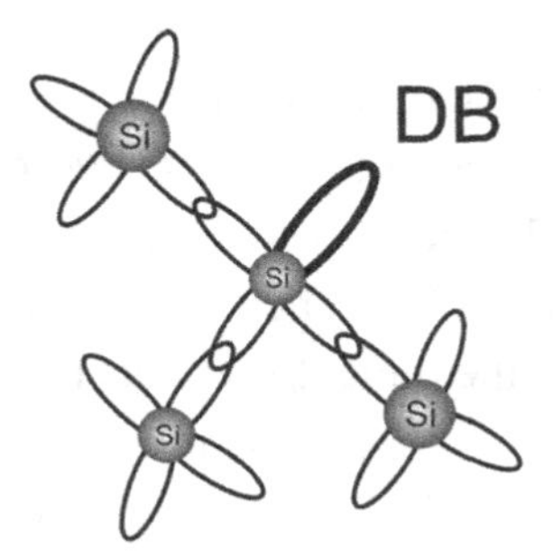

Figure 4.2. An sp^3 orbital remains as an unpaired and unsatisfied valence state. This dangling bond (DB) forms an electronic localized state.

It can be concluded from diffraction measurements that the upper limit for the first-neighbor distance in a-Si is about 0.27 nm. A special case occurs when only three neighboring atoms are situated within this distance. Three sp^3 orbitals overlap with the orbitals of neighbor atoms and form normal sigma bonds. The three bonds are usually referred to as "backbonds" and the three atoms as "backbond atoms." One sp^3 orbital remains as an unpaired and unsatisfied valence state, as shown in Figure 4.2. This is called the dangling bond (DB), and it is localized. This single electron has an uncompensated spin, which can be detected by electron spin resonance (ESR) measurements. Dangling bonds dominate electronic and optical properties and hence affect the characteristics of the device. Amorphous semiconductors with lesser defects (DBs, etc.) are often referred to as device-quality materials.

Exception: graphitic carbon

Let us consider a simple example. Ethylene (C_2H_4) has a double bond between the carbon atoms. For this molecule, the wave functions of the carbon atoms will be sp^2 hybrids, because one pi bond is required for the double bond between the two carbon atoms. Only three sigma bonds can be formed per carbon atom. In sp^2 hybridization, the $2s$ orbital is mixed with only two (p_x and p_y) of the three available $2p$ orbitals. These three hybrids are also orthonormalized states. The mathematical terms are as follows:

$$
\begin{aligned}
sp^2 &= \left(\frac{1}{3}\right)^{1/2} s + \left(\frac{2}{3}\right)^{1/2} p_x, \\
sp^2 &= \left(\frac{1}{3}\right)^{1/2} s - \left(\frac{1}{6}\right)^{1/2} p_x + \left(\frac{1}{2}\right)^{1/2} p_y, \\
sp^2 &= \left(\frac{1}{3}\right)^{1/2} s - \left(\frac{1}{6}\right)^{1/2} p_x - \left(\frac{1}{2}\right)^{1/2} p_y.
\end{aligned}
\tag{4.2}
$$

In ethylene the two carbon atoms form a sigma bond by overlapping two sp^2 orbitals, and each carbon atom forms two other covalent bonds with hydrogen via an s–sp^2 overlap, all with 120° angles. The pi bond between the carbon atoms perpendicular to the molecular plane is formed by the $2p_z - 2p_z$ overlap. The hydrogen–carbon bonds are all of equal strength and length (resonant bonds), which agrees with experimental data. In a honeycomb graphite crystal, such a local arrangement can be observed as planar. The bond length is shorter by 0.01 nm, and the bond is stronger than in a diamond crystal because of the additional pi bond. This local arrangement with sp^2 hybridization forms the graphite-like a-C.

Group VI elements

Selenium atoms contain six valence electrons ($4s^24p^4$), and the previously described sp^3 hybridization is no longer valid. From s, p_x, p_y, and p_z only four independent orbitals can be constructed, i.e. two out of four must be doubly occupied. A simple hydrogen selenide molecule has the atomic structure H–Se–H, with a bond angle of 91°. Note that in hydrogen telluride the bond angle is equal to 90°. As a first approximation, we can explain the bonding process in the following way. Two near-perpendicular sigma bonds can be constructed by putting one electron in the p_x and another in the p_y atomic orbitals, which are overlapped by the $1s$ orbital of hydrogen atoms. The other two unshared electrons occupy the p_z atomic orbital, forming a non-bonding

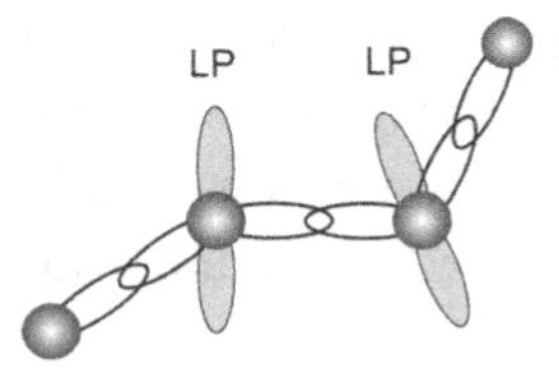

Figure 4.3. The p_x and p_y orbitals overlap with similar orbitals of adjacent atoms and form sigma bonds. Two unshared electrons occupy the p_z atomic orbital, forming a non-bonding lone-pair (LP) state.

lone-pair (LP) state. The other two electrons remain untouched in the s atomic orbital.

Pure non-crystalline selenium is the model material for chalcogenide glasses. In the condensed phase, selenium atoms form a chain-like structure. It can be concluded that only two electrons in the p_x and p_y orbitals form sigma bonds that overlap with similar orbitals of adjacent atoms (see Figure 4.3). Bond angles are slightly larger than 90°. This means that the s atomic orbital is not exactly untouched; a small contribution from the s orbital is required to form near-perpendicular sigma bonds and a LP. The energy of these non-bonding LP states lies between the electron-bonding and antibonding energy levels.

Selenium chains have unpaired and unsatisfied valence states (DBs) at their ends. Unlike a-Si:H, this localized state is not uncompensated, i.e. these orbitals are doubly occupied by electrons, so we have C_1^- , where the minus superscript refers to the atomic net charge and the subscript denotes the coordination number.

4.2 Electronic density of states

In crystalline materials the presence of long-range order and perfect translational symmetry greatly simplifies the mathematical treatment of the electrons. The electron states can be described by Bloch wave functions extending throughout the crystal (delocalization) as follows:

$$\varphi(\boldsymbol{k},\boldsymbol{r}) = u(\boldsymbol{k},\boldsymbol{r})\exp(i\boldsymbol{k}\boldsymbol{r}), \tag{4.3}$$

where $u(\boldsymbol{k}, \boldsymbol{r})$ describes the periodicity of the crystal lattice, i.e. $u(\boldsymbol{k}, \boldsymbol{r}) = u(\boldsymbol{k}, \boldsymbol{r} + \boldsymbol{R})$, where $\boldsymbol{R}$ is the lattice translation vector, and the exponential term represents a plane wave. The wave vectors $\boldsymbol{k}$ of the electrons are related to the translational symmetry of the lattice. It follows that a reciprocal lattice related to the unit cell parameters is established in k-space. This solid state physics tool is inadequate for non-crystalline materials. The lack of long-range order does not allow the wave vector $\boldsymbol{k}$ to be defined, and therefore the classification of electronic band structure $E(\boldsymbol{k})$ also cannot be applied. Furthermore, this prevents the traditional use of the definition of effective mass (Singh, 2002).

Instead of $E(\boldsymbol{k})$, the number of electronic states at energy E per unit energy $N(E)$ is used as a well-defined expression for amorphous materials. The occupancy is derived by a Fermi distribution function at a given temperature T, which is nearly a step function at room temperature. As the temperature rises above absolute zero, there is more energy to spend on lattice vibration and on lifting some electrons into an energy state of the conduction band (CB). Electrons excited to the conduction band leave behind holes in the valence band (VB). Both the CB electrons and the VB holes contribute to electrical conductivity.

The most important property of semiconductors is the existence of a bandgap which separates the valence and conduction bands. The presence of perfect periodicity in crystalline semiconductors helps to derive this bandgap. However, the definition of the bandgap is not as straightforward for amorphous semiconductors. The first pioneering attempt to define the bandgap was made by Weaire and Thorpe (1971) for tetrahedrally bonded amorphous semiconductors. In both crystalline and amorphous semiconductors, bands are desribed by a local arrangement. The chemical bonding can be described using a tight-binding (TB) model with the following Hamiltonian:

$$H = V_1 \sum |\theta_{ik}\rangle\langle\theta_{il}| + V_2 \sum |\theta_{ik}\rangle\langle\theta_{jk}|, \tag{4.4}$$

where θ_{ik} denotes the kth sp^3 hybrid orbitals associated with the ith atom. The first term is a sum over the interaction in which the kth and lth wave functions belong to the same atom, and the second term is a sum associated with the wave functions with the same bonds. These are the strongest interactions; this is why the Weaire–Thorpe model is a good starting approximation: $V_1/V_2 = 1/3$ for silicon. It can be shown that there is no gap for the case of $V_1/V_2 = 1/2$ only (i.e. the valence and conduction bands touch each other).

Another important law for amorphous semiconductors is Anderson's theory of localization (Anderson, 1958). A simple description of this law is that the increase in the disordered potential causes electron localization, i.e. the wave function becomes confined into a small volume. This characteristic difference between the electronic structure of crystalline and non-crystalline solids plays an essential role in the transport and optical properties, etc.

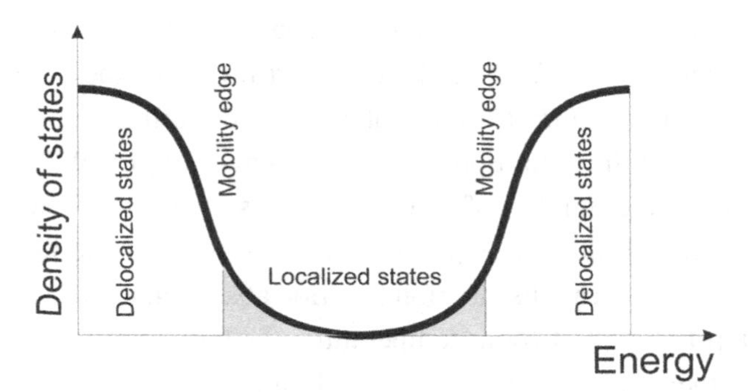

Figure 4.4. Electronic DOS of tetrahedrally bonded pure amorphous semiconductors.

The rapid growth in the number of computer facilities and faster computational speeds mean that there are many different numerical methods that can be used to calculate the electronic structure of condensed matter. The translational symmetry of crystalline materials is beneficial because of the application of periodic boundary conditions, therefore only a few atoms are needed in the numerical calculations to obtain an accurate result. For amorphous materials, a periodic boundary condition using so-called "super cells," is also applicable, although there must be a large number of randomly distributed atoms to produce reasonable results. Here, the term "large number" means at least 100 atoms. Another possible method is cluster calculation, in which a structural model, again containing a large number of atoms, is required. At the surface of these clusters, the DBs should be saturated, usually with hydrogen termination. A drawback of hydrogen termination is that a large charge transfer occurs in Si–H bonds. To resolve this problem, a longer Si–H bond length than the average normal Si–H atomic distance must be substituted into the model structure.

Three basic methods have been developed and used for the calculation of electronic properties of amorphous semiconductors: tight-binding models, Hartree–Fock methods, and density-functional theories. For the electronic DOS calculation, the Hartree–Fock approximation usually overestimates the energy gap, while the density-functional methods underestimates it.

The main conclusions of a large number of computer calculations and theoretical work are displayed in Figure 4.4. In the DOS of a fourfold-coordinated pure amorphous semiconductor, delocalized and localized regions (tail states) can be found separately by the mobility edge. Such localized band tails do not exist in crystalline counterparts. Following Fermi statistics, the lower-energy

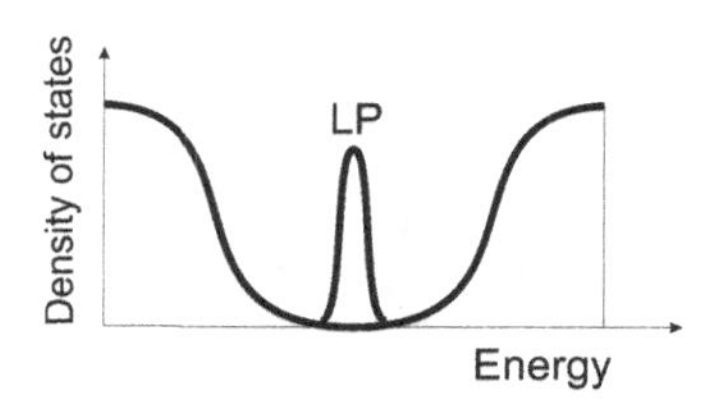

Figure 4.5. Non-bonding LP states are in the middle of the gap in the electronic DOS.

region is almost completely filled by electrons as a function of temperature. The pure chalcogenide glasses, however, have a different electronic DOS. As shown in Figure 4.5, an extra peak appears in the gap; the so-called non-bonding LP states (bands) are in the middle of the gap between the bonding and antibonding states (bands). The Fermi energy, $E_F = \mu(T = 0)$ lies between the top of the LP band and the conduction band. This characteristic feature in the DOS of chalcogenides plays a primary role in photoinduced effects, which will be discussed in Chapter 5.

4.3 Defects

Classification of defects

One possible definition of a defect is something that prohibits or prevents perfection. Structures that are devoid of defects are considered *ideal*. Inclusion of a structural defect thus refers to a configuration in which an atom, or group of atoms, does not satisfy the structural rules belonging to the ideal reference state of the material. Real materials contain structural defects that can dominate their physical and chemical behaviors. In covalently bonded semiconductors, defects usually create states inside the energy gap which control the electronic and optical properties. Some structural defects can capture electrons or holes. It is not easy to identify structural defects in diffraction measurements. The key defects that occur in crystalline materials comprise the following:

- point defects due to impurity atoms in either substitutional or interstitial sites;
- complexities associated with special correlations between different point defects such as impurity–vacancy pairs or donor–acceptor pairs;
- one-dimensional (line) defects, including translational displacements of atoms (dislocations) and rotational displacements (disclinations);

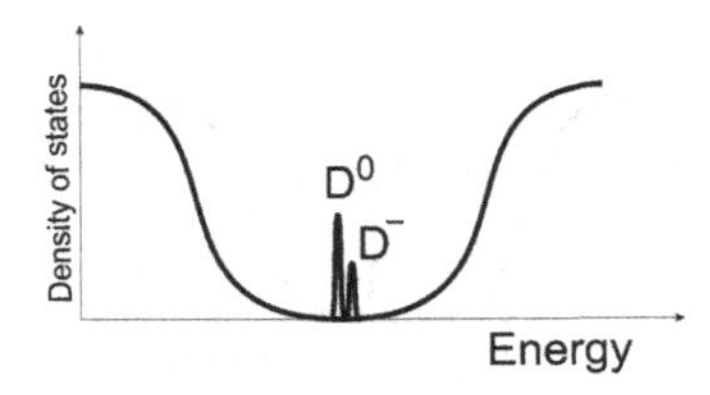

Figure 4.6. Defect model in a-Si. D^0 (D^+) and D^- indicate neutral (positively charged) and negatively charged dangling bonds. D^+ and D^- have no spin.

- two-dimensional (planar) defects, including low-angle grain boundaries;
- native point defects, such as vacancies and interstitial atoms;
- defects associated with grain boundaries in polycrystalline materials;
- distortions caused by surfaces or interfaces; and
- voids, i.e. several connecting atoms are missing from the network.

What is the meaning of defects in amorphous semiconductors? At first, this may seem to be a silly question because amorphous means disorder and so there is obviously a lack of perfection; however, this statement is incorrect because the disorder in amorphous semiconductors has order in some sense. As an example, consider a-Si, in which all the atoms are tetrahedrally coordinated, similar to the crystalline counterpart. However, there is a lack of long-range order. In this case, "defects" can be regarded as the deviation from the ideal picture. In this book, we define a "defect" as a dangling bond, as discussed in Section 4.2. There are two different classes of defects: natural and photo-created defects. In this chapter, we discuss natural defects.

In tetrahedrally bonded semiconductors broken bonds (unsaturated bonds) are observed, as shown in Figure 4.2, that break the local topological order. These dangling bonds (DBs) release the internal tension inside the continuous random network. A simple DB normally contains one electron and is electrically neutral, but under certain circumstances (chemical or electronic doping) the electronic occupancy changes from D^0 (neutral) to D^+ (positive) and D^- (negative), where D denotes the DBs, as shown in Figure 4.6. As D^- has an extra electron, the energy level is raised by an amount U_c (the Coulombic repulsive energy at the DB). Note, however, that D^+ should have the same energy level as D^0; D^- does not have spin. To reduce the number of DBs, the amorphous silicon (and germanium) are passivated by hydrogen, which reduces the DB density by several orders of magnitude. These intrinsic defects form electronic states inside the gap, and this under-coordination defect has a significant ESR signal with $g = 2.0055$ in a-Si, which is the most important experimental evidence of DBs. The spin density is estimated to be around 10^{19} cm^{-3} in a-Si, but is much less (around 10^{15} cm^{-3}) in a-Si:H, which results from the preparation techniques and the hydrogen contamination. Such intrinsic defects do not exist in crystalline Si because removing one atom produces a vacancy with four DBs. The energy of this arrangement is too high, and bond reconstructions occur. Pantelides (1986) has suggested other coordination defects, namely over-coordinated atoms (floating bonds). Whether floating bonds exist or not is still not clear. If we consider the position of a fourth-nearest-neighbor fourfold-coordinated atom with respect to that of a threefold-coordinated atom, along with the direction of the DBs with respect to this atom, we can say that a weak interaction probably exists between the two atoms. This geometrical arrangement provides a fivefold-coordinated atom. However, the weak interaction does not contain two localized electrons; it is not a normal sigma bond, but it can be considered to be a bond in some sense.

Theoretically, DBs can also consititute a possible defect in chalcogenide glasses. Group VI elements in the amorphous phase have long chains and rings. At the ends of the chains are onefold-coordinated atoms, and unpaired spins are expected. Experiments show different results compared to those with a-Si. Pure a-Se presents no ESR signal in dark conditions, i.e. there is no neutral DB with an unpaired spin. Originally, Anderson (1975) proposed a negative-U concept, and then this idea was applied to chalcogenide glasses (see, for example, Mott and Davis 1979). Normally, a Coulomb repulsive interaction should occur between two electrons. In the Anderson theory, the phonon–electron interaction effectively gives rise to an attraction. The Hamiltonian describing the phenomenon has three parts: electronic, phonon, and electron–phonon. If the coupling constant in the electron–phonon part is large enough, the effective correlation between electrons is negative (a negative-U center) and, instead of C^0, two other charged states, C^+ and C^-, can be formed, in which there are no ESR-active spins:

$$2C^0 \rightarrow C^+ + C^-. \tag{4.5}$$

This means that the charged states C^+ and C^- are more stable than the neutral state C^0. Kastner, Adler, and Fritzsche (1976) showed that these are threefold (C_3^+) and onefold (C_1^-) coordinated in terms of the chemical bonds, and are called the valence-alternation pairs (VAPs). Furthermore, another defect, the threefold-coordinated Se atom (junction), can be formed in the network.

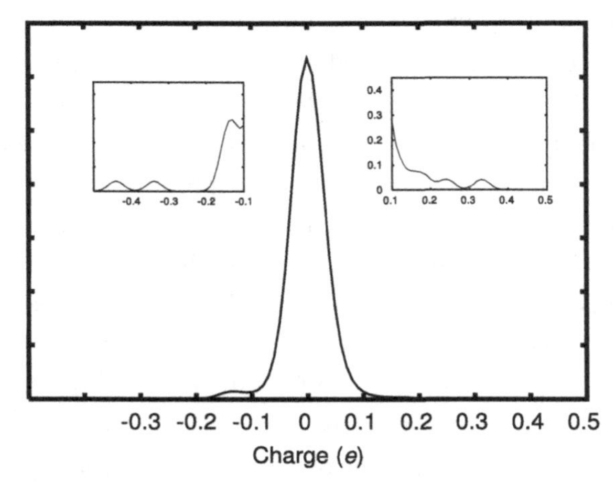

Figure 4.7. Charge distribution of a 162 Se atom cluster calculated using the DFT GGA method. Larger charge accumulations can be observed in both directions (see insets). (From Lukács, Hegedus, and Kugler (2009). *J. Mater. Sci.: Mater. Electron.*, **20**, S33–S37. With kind permission from Springer Science and Business Media.)

A controversial consequence of the Anderson theory has been reported by Lukács, Hegedus, and Kugler (2009). The charge distribution was calculated in a-Se using two different calculation methods: a density-functional method and a tight-binding (TB) model. Figure 4.7 displays the charge distributions using the DFT GGA (generalized gradient approximations) calculation on 162 atom clusters. Essentially, the absolute values of the charge accumulations on the twofold-coordinated atoms are less than $0.1e$. Larger charge accumulations in both positive and negative directions were observed (see the insets in Figure 4.7). The larger positive charge accumulation belongs to the threefold-coordinated atoms, and the negatively charged atoms are onefold coordinated. A similar effect was observed using the TB model. The charges in the TB calculations are somewhat larger than those calculated using DFT (see Figures 4.7 and 4.8). These calculations do not contain any phonon–electron interactions, and the results suggest that the threefold-coordinated atoms (which look similar to VAPs) lose an electron, and that these electrons are transferred to the end of chains. However, these are not negative-U defects caused by phonon–electron interactions.

In amorphous semiconductor networks, coordinated impurity atoms having no unsatisfied bonds are usually electronically neutral. The most important conclusion of Mott's famous $(8 - N)$ rule (1969) is that amorphous semiconductors

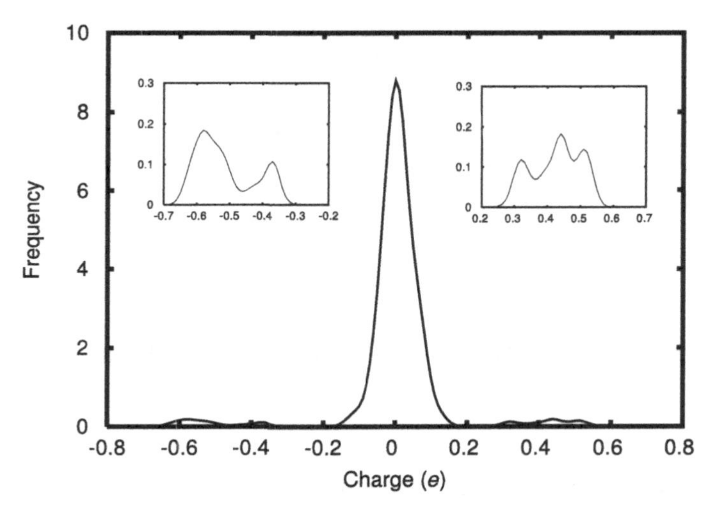

Figure 4.8. Charge distribution calculated using a TB model on the same Se cluster as in Figure 4.7. (From Lukács, Hegedus, and Kugler (2009). *J. Mater. Sci.: Mater. Electron.*, **20**, S33–S37. With kind permission from Springer Science and Business Media.)

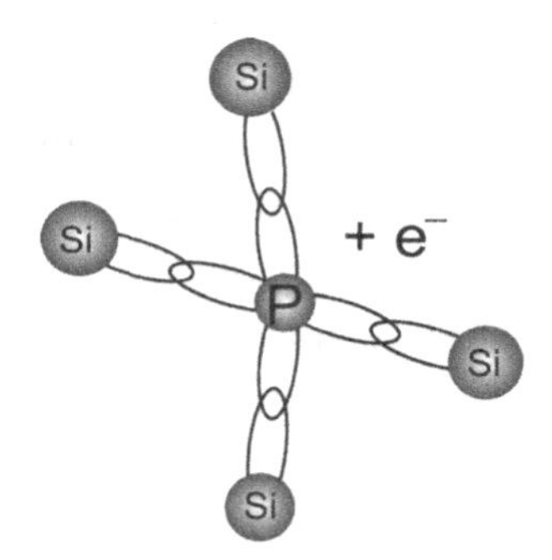

Figure 4.9. p doping: schematic representation of spatial distribution of electron excess around a phosphorus atom in a silicon network.

cannot be doped! Some years later, Spear and LeComber (1975) showed that plasma-glow-discharge deposition of silicon, using mixtures of silane and either phosphine or diborane, enabled electrically active pentavalent (P) (see Figure 4.9) and trivalent (B) (see Figure 4.10) impurities to be incorporated into the films, making them n type and p type, respectively. These impurities introduce unsatisfied bonds into the network that are electronically active. An interesting historical fact is that the first reported, but not successfully exploited, experiment on doping had been carried out earlier by Chittick, Alexander, and

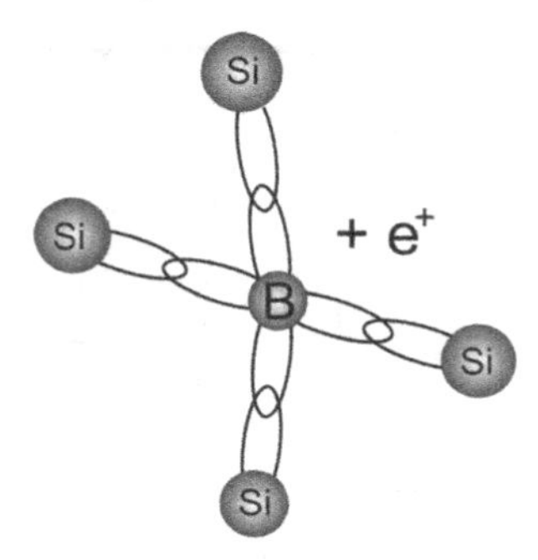

Figure 4.10. n doping: schematic representation of spatial distribution of electron deficiency around a boron atom in a silicon network.

Sterling (1969) at the Standard Telephone Laboratories in Harlow, UK. They stated in the abstract of their paper that "the effects of heat-treatment, ageing, and doping on the properties of amorphous silicon are reported."

Estimation of the doping energy levels for a crystalline arrangement is quite an easy task. A hydrogen molecule type model is a good approach to take (see, for example, Sólyom, 2009). The core charge of a phosphorus atom is higher than that of the connecting atom, and the extra electron is weakly bound to the phosphorus atom by the Coulomb field. For the calculation it must be considered that a free electron moves in a dielectric media instead of a vacuum and that the electron inside a periodic potential has an effective mass. Mostly, in heavy-doped semiconductors two doping atoms can be located nearby, i.e. in first-, second-, third-, etc. neighbor positions, and the interaction between them is important. A hydrogen molecule type model should be used in the energy-level calculation for heavy-doped semiconductors.

In non-crystalline semiconductors the lack of a well-defined effective mass means that this hydrogen atom/molecule type model is useless for such materials. The position of the energy levels belonging to dopant pairs was investigated in fourfold-coordinated a-C (ta-C) and in a-Si by means of a Hartree–Fock *ab initio* method (Kádas and Kugler, 1997a,b; Kádas, Ferenczy, and Kugler, 1998). Models contained 45 to 583 carbon or silicon atoms, and boron, phosphorus, and nitrogen impurities were incorporated into the amorphous networks. It has been observed that the positions of the midgap states are primarily determined by the separation of the impurity atoms. The electronic density of states in ta-C doped with two nitrogen atoms is shown in Figure 4.11. It should be noted that the Hartree–Fock method always overestimates the gap energy. Peaks 1 and 2

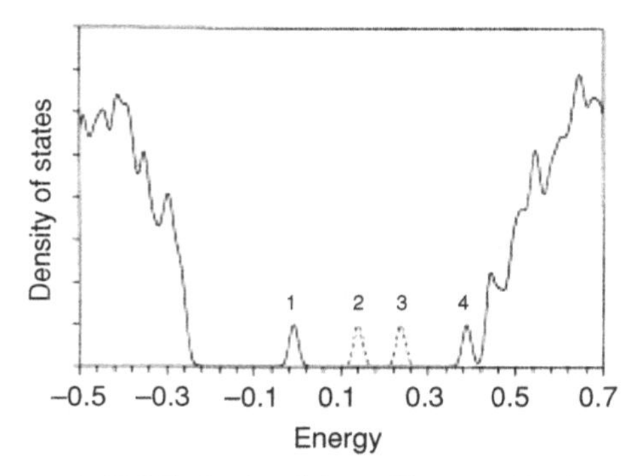

Figure 4.11. Electronic DOS obtained from Hartree–Fock *ab initio* cluster calculations for different ta-C models doped by two nitrogen atoms. (From Kádas, Ferenczy, and Kugler (1998). *J. Non-Cryst. Solids*, **227–230**, 367. Copyright 2013 with permission from Elsevier.)

display the highest occupied molecular energies when two nitrogen atoms are at the first- and fourth-neighbor positions, respectively. Peak 3 belongs to a ta-C cluster containing only one nitrogen dopant. This latter configuration occurs when the other dopant is placed at an infinite distance. These direct calculations of the energy levels suggest that a significant difference can occur between energy levels as a function of doping rate.

The diamond crystal structure consists of sixfold atomic rings, whereas the corresponding amorphous diamond-like structure is made of five-, six-, and sevenfold rings, with the threefold and fourfold rings being atypical. Nevertheless, triangle and squares break the local order, i.e. they can be considered as one type of defect. Such defects were found in a-Si (Kugler *et al.*, 2003). As a consequence, two characteristic peaks can be found inside the tail of the band structure (Kugler, 2012). The first larger peak corresponds to the square atomic arrangements, whereas the peak at the higher energy is due to triangular arrangements (Figure 4.12). In earlier investigations of electron transport, hopping conductivity, optical properties, etc., the tail was usually considered to be an exponential or Gaussian decaying function. Triangles and/or squares have never been considered in any band structure calculations, although they play an important role in several phenomena.

Normally a two-component (A and B) covalently bonded amorphous network contains A–A, A–B, and B–B bonds. In special stoichiometric compositions ($GeSe_2$ or As_2Se_3, etc.) only A–B bonds are energetically favorable. In these chalcogenide materials, a few A–A and B–B bonds can be found

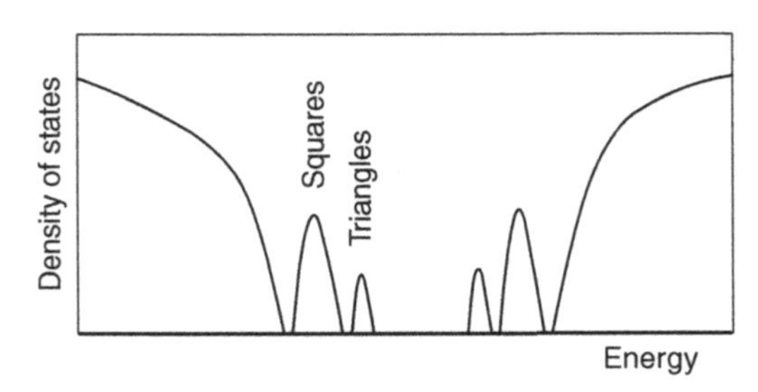

Figure 4.12. Electronic density of states of amorphous semiconductors containing two significant peaks belonging to the triangle and square defects.

experimentally. This class of defects is known as "wrong bonds." Various authors have worked on this problem, among them Petri, Salmon, and Fischer (2000), who published an excellent paper on the experimental evidence for wrong bonds. Neutron-diffraction measurements were made in the D4B instrument at ILL Laue-Langevin, Grenoble, with $GeSe_2$ as the target material. A significant number of Ge–Ge or Se–Se homopolar bonds were identified, which demonstrates the existence of wrong bonds.

In amorphous networks, voids where several connecting atoms are missing from the network also occur as defects. The properties and number of these defects strongly depend on the preparation technique. In obliquely deposited amorphous $GeSe_3$ films, the density decreased when the angle of the evaporate beam increased, as reported by Rayment and Elliott (1983). Columnar structures of these materials were also observed in obliquely deposited a-GeS_2. A decrease in both the refractive index and the microhardness versus angle of incidence has been reported. These results provide evidence for increasing free volume with increasing obliqueness (Starbova *et al.*, 1999). Atomic-scale computer simulations for structures of flatly and obliquely deposited amorphous selenium films have been carried out to understand how the properties of chalcogenide glasses are influenced by their preparation method (Lukács, Hegedus, and Kugler, 2008). It was concluded from theoretical structure work that the obliquely deposited a-Se thin films contain more coordination defects, lower densities, and larger voids than the flatly deposited ones inside the networks.

Finally, an interesting topological defect is discussed for a-C, which has different percentages of fourfold- and threefold-coordinated atoms. The pi–electron network is determined by the threefold-coordinated graphite-like atoms. These atoms form three localized sigma bonds plus a delocalized pi bond. Graph theory can be used to estimate the number of topologically derived defects in graphite-like a-C (graphene) in relation to the electronic DOS at the Fermi level (Kugler and Laszlo, 1989). The number of topologically derived energy levels is determined by the global topological properties of the graphite-like a-C structures (Laszlo, 1993, 2000).

Defect spectroscopy

Defects produce *localized* states in the energy gap, and hence they can be detected by different spectroscopic techniques. Since the defects dominate the electronic and optical properties, it is important to estimate the number of defects and their energy levels *experimentally*. Optical, electrical, and magnetic measurements are the principal techniques used in defect spectroscopy, which will be briefly reviewed in this section. The details of defect spectroscopy in a-Si:H and amorphous chalcogenides (a-Chs) are described elsewhere (Singh and Shimakawa, 2003).

Optical measurements

The optical absorption (the so-called "midgap absorption") related to the defect is usually weak ($\alpha \leq 10\ \text{cm}^{-1}$) in a-Si:H and a-Chs, and hence several sensitive techniques are adopted, as described in the following.

(i) Photoinduced absorption (PA) monitors variations in the intensity of probe light absorbed by trapped carriers in defects, after the carriers are created in the extended states by illumination. Owing to the increase in absorption, the transmittance of the probe light changes. The spin-dependent PA (accompanying electron spin resonance) is a sensitive technique (Hirabayashi and Morigaki, 1983).
(ii) Photothermal deflection spectroscopy (PDS) measures the heat absorbed in the sample and is used to measure the lower optical coefficient of thin films (Jackson and Amer, 1982). A sample is immersed in liquid CCl_4. As non-radiative recombination is a dominant channel near room temperature, heat is generated by phonon emission. This heat generation, with an intensity-modulation beam of pumping light, induces a change in the refractive index of CCl_4, which is detected by the deflection of a laser beam passing through the sample. A He–Ne laser is usually installed to emit the probe light, and the periodic deflection is measured with a position sensor. If the wavelength

of the pump beam is varied, the deflection of the probe beam becomes a measure of the optical absorption spectrum of the materials. Some calculations, however, are required to obtain the absolute optical absorption coefficient. The PDS sensitivity is very high; for example, for a 1000 nm (1 μm) thickness, values of α as low as 0.1 cm^{-1} can be obtained.

Electrical measurements

The recombination and emission of carriers at defects are investigated by electronic transport. As the thermal emission rate is related to the energy level, and the shift of the Fermi level is related to the number of trapped carriers, these two parameters are estimated in principle from the electrical transport measurements. Several techniques are described in the following.

(i) Field-effect measurements were first applied to obtain the localized density of states in a-Si:H (Madan, LeComber, and Spear, 1976) and in a-Chs (Marshall and Owen, 1976). A voltage is applied across the dielectric thin layer deposited on the materials that yield band bending, producing a space charge near the interface. Note that this is the same principle as for a thin-film transistor. This produces an excess of carriers in the extended states (excess conduction), which is due to the shift of the Fermi level toward the band edge. The DOS near the Fermi level is estimated in this method. As most of the space charges are 10 nm from the interface, they are strongly affected by the presence of interface states, which may be a drawback of this technique.

(ii) Capacitance measurements may be made by considering the depletion layer of a Schottky barrier under reverse bias. As described in standard text books, capacitance varies with the bias voltage, reflecting the space-charge density, and the space charge (i.e. the localized carriers residing at the defects) can be estimated from the capacitance value. The most familiar technique that is used to measure deep levels in semiconductors is deep-level transient spectroscopy (DLTS) in which the transient capacitance of a Schottky barrier is measured (Lang, 1974). In DLTS, all the traps are filled by applying a forward bias to the Schottky barrier, and subsequently a reverse bias. The depletion layer width decreases over time to its steady state value. Note that the capacitance is inversely proportional to the thickness of the depletion layer.

(iii) The space-charge-limited current (SCLC) technique measures the current–voltage (I–V) characteristics. Deep traps can be filled by current injection, and hence the Fermi level shifts toward the relevant band states (the conduction band for electrons and the valence band for holes). There are several ways of analyzing the I–V characteristics (Mackenzie, LeComber, and Spear, 1982, Shimakawa and Katsuma, 1986). Although SCLC is a simple technique, good injected electrodes (contacts) are not easy to prepare. If the number of defects is large, the shift of the Fermi level is small, and hence the covered energy range of the DOS becomes narrow.

(iv) The constant-photocurrent method (CPM) monitors the illumination intensity G, keeping the photocurrent constant, as a function of photon energy (Vanecek *et al.*, 1983). In this method, the quasi-Fermi level is kept the same, independent of photon energy, and hence the recombination time is unchanged. Because of this condition, it is possible to deduce the optical absorption coefficient α (around 0.1 cm^{-1}). This technique has been applied to both a-Si:H and a-Chs, and yields information on the weak optical absorption tails (see, for example, Singh and Shimakawa, 2003).

(v) AC loss occurs during: (a) electric dipole motion, (b) hopping of carriers between localized states (a-Chs), and (c) inhomogeneous media (a-Si:H). These will be discussed in detail in Section 4.4. Via the analysis of ac loss, for example in a-Chs, an acceptable density of charged defects (10^{17}–10^{18} cm^{-3}) of reasonable energy at a location in the bandgap can be deduced (Ganjoo and Shimakawa, 1994).

(vi) The electrophotography (or xerography) method also provides the DOS in a-Si:H (Imagawa, Akiyama and Shimakawa, 1984) and in a-Se (Abkowitz and Markovics, 1984). This method is a relatively new candidate with which to evaluate the DOS; it is able to ignore the surface and interface effects, which cannot be neglected in the FE and capacitance–voltage (CV) methods. The principle is the same as that employed in commercial xerographic equipment.

Magnetic resonance

A quantum state, for example a defect state, occupied by a single electron is split by a magnetic field; this is known as Zeeman splitting.

(i) Electron spin resonance occurs due to a transition of electrons between the split energy levels. The transition occurs at microwave frequencies for a conventional magnetic field (~3 T) between two Zeeman levels with unpaired electrons whose spin states $S = 1/2$ (i.e. the paramagnetic state). ESR signals can be observed by probing the microwave absorption if the defects have unpaired spins larger than 10^{15} cm^{-3}, which of course depends on the sensitivity of the equipment. A DB in a-Si:H has unpaired spin, and in the so-called "device-quality" a-Si:H film the spin density is less than 10^{16} cm^{-3} (Street, 1991). When an unpaired electron interacts with nuclear spin, the hyperfine structures are observed in the ESR spectra. Thus the ESR spectra contain the information of the local configuration of these spin states.

ESR signals in pure a-Chs are generally not observed under dark conditions, because, unlike for a-Si:H, only the charged defect states (no unpaired spins) are expected to exist in a-Chs (except for some components). Under photoillumination at lower temperatures, however, ESR signals appear; we call this light-induced ESR (LESR) (Bishop, Strom, and Taylor, 1975). Note that LESR signals appear in a-Si:H at low temperature, which is not the case for the original ESR signal under dark conditions (Street, 1991). Various photoinduced effects have been monitored in experiments using LESR (Shimakawa, Kolobov, and Elliott, 1995). A pulsed ESR technique, in which transient signals produced by coherent excitation using microwave pulses are detected in the time domain, is also useful (Isoya *et al.*, 1993).

(ii) Electron nuclear double resonance (ENDOR) is helpful in obtaining structural information about defects and their surroundings when hyperfine interactions are weak. Measurements using ENDOR monitor the variation of the ESR with nuclear magnetic resonance through hyperfine interaction. The intensity of the ESR signal changes when NMR occurs. The details of ENDOR measurements on a-Si:H have been described elsewhere (Morigaki, 1999).

4.4 Optical and electronic properties

Fundamental optical absorption

The DOS for both conduction and valence bands is expected to be proportional to the square root of the energy in 3D materials. It should be noted that the wave vector $\boldsymbol{k}$ is not a useful quantum number and therefore it cannot be applied in amorphous solids. Thus the terms used to describe direct and indirect optical transitions between valence and conduction bands in the study of crystalline solids are not useful when considering amorphous materials. Furthermore, the DOS in disordered matter, in general, may be described by taking into account the *fractal concepts* (Mandelbrot, 1982) that are known to dominate many physical properties in amorphous semiconductors (Zallen, 1983).

Based on fractal geometry, i.e. the spherical volume V is proportional to r^D, where r is the radius and D is the fractal dimension, which is different from the values defined by the Euclidean space dimension d (= 1, 2, 3), the DOS is given by

$$N(E)\,dE = AE^{(D-2)/2}dE, \tag{4.6}$$

where A is a constant (He, 1990). In a normal homogeneous space, D should be the same as the Euclidean space dimension d. For example, $D = d = 3$ leads to $N(E)$ proportional to $E^{1/2}$, as described in any standard text book. When a material contains a large amount of voids, D can be significantly lower than the dimensions of the Euclidean space.

Let us discuss the fundamental optical absorption in fractal space with dimension D. For interband electronic transitions, the optical absorption coefficient can be written as follows (Mott and Davis, 1979):

$$\alpha(\omega) = B\int \frac{N_c(E)\,N_v(E - \hbar\omega)\,dE}{\hbar\omega}, \tag{4.7}$$

where B is a constant which includes the square of the transition matrix element as a factor, and the integration is over all pairs of states in the conduction $N_c(E)$ and valence $N_v(E)$ bands. Let us express the DOS of the conduction and valence bands as

$$N_c(E) = const\,(E - E_c)^{\alpha}, \tag{4.8}$$

$$N_v(E) = const\,(E - E_v)^{\beta}, \tag{4.9}$$

where $\alpha = (D_c - 2)/2$ and $\beta = (D_v - 2)/2$. Here D_c and D_v are the fractal dimensionalities of the conduction and valence bands, respectively. Then eqn. (4.8) produces (Nessa *et al.*, 2000)

$$\alpha(\omega)\hbar\omega = B'(\hbar\omega - E_o)^{\alpha+\beta+1}, \tag{4.10}$$

where B' is another corresponding constant. This yields

$$[\alpha(\omega)\hbar\omega]^n = B'^{1/n}(\hbar\omega - E_o), \tag{4.11}$$

where $1/n = \alpha + \beta + 1$. If the form of both $N_c(E)$ and $N_v(E)$ is parabolic, i.e. $\alpha = \beta = 1/2$ for 3D materials as in the most of the studies, then eqn. (4.11) becomes

$$[\alpha(\omega)\hbar\omega]^{1/2} = B(\hbar\omega - E_o), \qquad (4.12)$$

which is the well-known Tauc relation.

Tauc's relation for fundamental optical absorption is applied to many amorphous semiconductors to deduce the *optical gap*, E_o. There are many exceptions, for example n deviates from $1/2$ to take values between 0.3 and 1. Note that the valence band (VB) of a-Chs is formed by lone-pair (LP) electrons (known as LP semiconductors), and the conduction band (CB) is created by antibonding states. Thus the energy-dependent DOS form for the CB is expected to be basically different from that of the VB. A typical example may be $n \sim 1$ for a-Se. Although a LP band (VB) may have a 3D nature, the CB (antibonding states) should be characterized by a chain-like 1D nature, as long as no interaction between the chains is assumed. We therefore expect that $\alpha = -1/2$ and $\beta = 1/2$, producing $n = 1$ (Nessa *et al.*, 2000).

As we have already discussed, the energy dependence of the DOS for the CB and VB is expected to be different in a-Chs because the space dimensionality for the VB is larger than that for the CB. Therefore, we introduce D_c and D_v separately, and then n takes non-integer values for fractional-dimensional systems; $1/n = \alpha + \beta + 1 = (D_c + D_v - 2)/2$. In some a-Ch films, it is observed that D_c and D_v depend on the preparation conditions (Shimakawa, Singh, and O'Leary, 2006; Singh and Shimakawa, 2003).

Optical absorption tail (Urbach tail)

Following the fundamental optical absorption, an optical transition in the (exponential) localized tail states (lower-energy region) is observed. The origin of the exponential tail states is still a matter of debate, and hence there are many models available that discuss the origin of such localized tails (see, for example, Singh and Shimakawa, 2003; Tanaka and Shimakawa, 2011). In earlier investigations of electronic properties and related effects, the tail was usually considered to be an exponential or Gaussian decaying function. New types of defects, such as triangles and squares, have electronic states in the gap. If we accept this result, we see that the localized states have structure, i.e. two peaks can be found inside

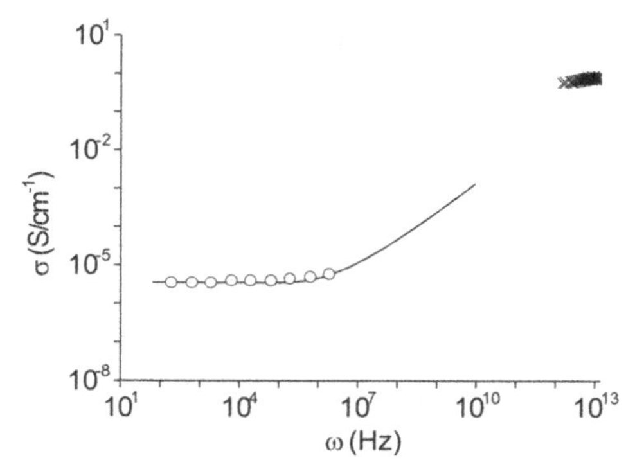

Figure 4.13. Frequency-dependent conductivity (real part) in a-Si:H. Open circles and crosses show the experimental data in the radio-frequency (rf) (Shimakawa and Ganjoo, 2002) and THz ranges (Fekete *et al.*, 2009), respectively. The solid line is a prediction using Dyre's expression (Dyre, 1988; Dyre and Schroder, 2000).

the tail (Kugler, 2012). The larger peak results from the squares, and the peak with larger energy is formed by triangles.

Photocarrier thermalization in picosecond time range and terahertz spectroscopy

The methods classically used to investigate photoconductivity are time-of-flight (TOF) spectroscopy (called *primary* photoconductivity) and the more common steady-state photoconductivity spectroscopy (called *secondary* photoconductivity). The TOF experiment, which is completely different from the secondary photoconductivity experiment, involves photocarriers, far from thermal equilibrium, that transit from the illuminated electrode to the counter electrode. As the details of photoconductivity spectroscopy have been reported extensively in the literature (for example, Street 1991; Morigaki, 1999, Singh and Shimakawa, 2003), we discuss only photocarrier transport with a subpicosecond time resolution (THz spectra) in a-Si:H. This work has been intensively studied and has presented excellent information on short-lived photocarriers in the CB (Fekete *et al.*, 2009; Lloyd-Hughes and Tae-in Jeon, 2012).

Figure 4.13 shows the frequency-dependent conductivity (real part) in a-Si:H. Open circles and crosses show the experimental data in the radio-frequency (rf) (Shimakawa and Ganjoo, 2002) and terahertz (THz) ranges (Fekete *et al.*, 2009), respectively. The solid line is a prediction using Dyre's expression of the ac loss

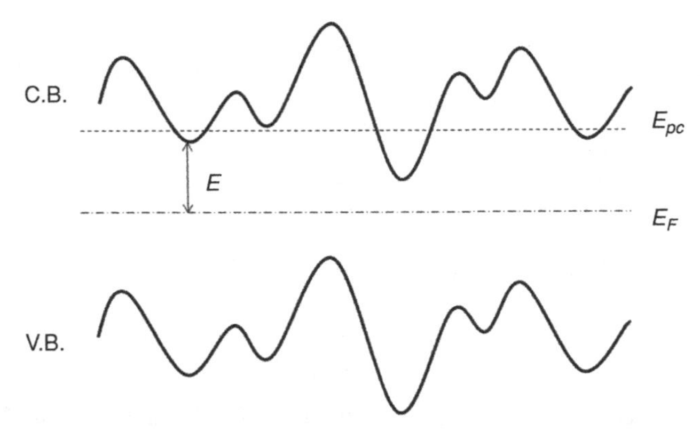

Figure 4.14. Potential fluctuation model in a-Si:H (Overhof and Thomas, 1989; Shimakawa and Ganjoo, 2002).

induced in inhomogeneous media (Dyre, 1988; Dyre and Schroder, 2000). Both the rf and THz ranges will be discussed in the following.

Usually, frequency-dependent conductivity (ac conductivity or ac loss) is considered to be caused by a *localized* motion of carriers (Mott and Davis, 1979; Singh and Shimakawa, 2003). A similar ac loss is also induced in macroscopically inhomogeneous media (Dyre and Schroder, 2000; Singh and Shimakawa, 2003). In the following, we briefly summarize the ac loss in inhomogeneous media, as this mechanism may dominate ac losses of a-Si:H in all frequency ranges.

A very simple percolation-path method proposed by Dyre (1988) was originally applied to the random walk (hopping) of localized carriers; now it is also applied to macroscopically inhomogeneous media. In inhomogeneous media, "space charges" relax with a relaxation time $\tau = \varepsilon_0\varepsilon_\infty/\sigma$, where ε_0 is the electric constant, ε_∞ is the background dielectric constant, and σ is the electrical conductivity. A space-charge relaxation induces the complex frequency-dependent conductivity as (Dyre, 1988)

$$\sigma^*(\omega) = \frac{c}{6}\left(-\mathrm{i}\omega + \frac{\mathrm{i}\omega\ln\left(\tau_{max}/\tau_{min}\right)}{\mathrm{In}\left[\left(1+\mathrm{i}\omega\tau_{max}\right)/\left(1+\mathrm{i}\omega\tau_{min}\right)\right]}\right) \equiv \sigma_R + \mathrm{i}\sigma_I, \quad (4.13)$$

where c is a constant and two cut-offs in the distribution of τ, $\tau_{max} = \varepsilon_0\varepsilon_\infty/\sigma_{min}$ and $\tau_{min} = \varepsilon_0\varepsilon_\infty/\sigma_{max}$, are introduced. Here, σ_{min} and σ_{max} are the local-minimum and local-maximum conductivities in the inhomogeneous media. Potential fluctuations, shown schematically in Figure 4.14 for an example of

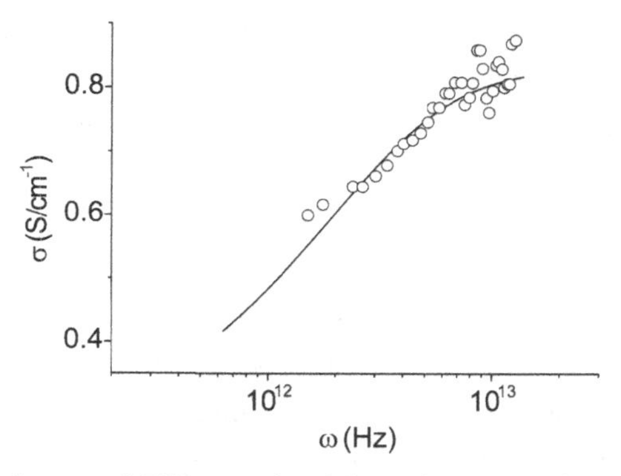

Figure 4.15. Real part of THz conductivity. The open circles and the solid line show the experimental data and the theoretical prediction, respectively. (From Shimakawa, Wagner, and Frumar (2013). *Phys. Status Solidi B*, **250**, 1004–1007. Copyright 2013 by John Wiley and Sons Inc.)

inhomogeneity, are suggested in a-Si:H (Overhof and Thomas, 1989). The local electronic conductivity therefore takes a variety of random values. Using

$$\ln(a + \mathrm{i}b) = \ln\sqrt{a^2 + b^2} + \mathrm{i}\tan^{-1}(b/a) \tag{4.14}$$

in the limit of $\omega \to 0$, we obtain

$$\sigma(0) = \frac{c\ln(\tau_{max}/\tau_{min})}{6\tau_{max}}. \tag{4.15}$$

Then we get the real and imaginary parts of the conductivity, $\sigma_R(\omega)$ and $\sigma_{I(}\omega)$, respectively. Note that c in eqn. (4.15) is a constant and that τ_{min} determines the high-frequency saturation of the conductivity.

The solid line in Figure 4.13 is the predicted real part of the conductivity, σ_R, obtained from $\sigma(0) = 3.5 \times 10^{-6}$ S cm^{-1}, $\tau_{max} = 2 \times 10^{-6}$ s. It is not easy to predict the value of τ_{min}; the predicted $\sigma(\omega)$ is not plotted above $\sim 1 \times 10^{10}$ Hz. Terahertz conductivities, shown by crosses ($\times$) in Figure 4.13, are obtained from the optical pumping THz spectroscopy (Fekete *et al.*, 2009), and these are very much higher than those reported in the rf range. This is due to optical pumping, which produces a large number of free carriers around 1×10^{18} cm^{-3} just after photoexcitation (Fekete *et al.*, 2009). We will discuss THz conductivity in more detail.

The open circles in Figure 4.15 show the experimental data of the real part of the THz conductivity σ_R, as already shown in Figure 4.13. The solid line displays the predicted values using eqs. (4.13) and (4.15), with $\sigma(0) = 0.35$ S cm^{-1},

$\tau_{max} = 1.0 \times 10^{-11}$ s and $\tau_{min} = 4.3 \times 10^{-13}$ s. The fitting of the predicted data to the experimental data is good. Note that $\sigma(0) = 0.35$ S cm^{-1} does not refer to the actual dc conductivity. Rather, it can be regarded as the *virtual dc conductivity*, since, in the optical pumping technique, dc conductivity is not a meaningful concept.

We now know that the value of τ_{min} lies in the region of 10^{-13} s. Note again that τ_{min} is not deduced from the rf-range spectroscopy. Let us discuss what τ_{min} means. The ac loss in the rf range is interpreted under the assumption of potential fluctuations (Shimakawa and Ganjoo, 2002). Here, the saturated value of $\sigma(\omega)$ approaches 1 S cm^{-1}, and then the minimum dielectric relaxation time τ_{min} is expected to be 9×10^{-13} s. This value is close to that (4.3×10^{-13} s) deduced from experimental data. Thus, it is suggested that potential fluctuations which affect the rf conductivity also dominate conduction loss in the THz range (Shimakawa, Wagner, and Frumar, 2013).

An alternative cause of THz conductivity – carrier hopping – which may induce behavior similar to that previously described, should also be discussed. The inverse of τ_{min} (2.5×10^{12} s^{-1}) is close to the phonon frequency, and Fekete *et al.* (2009) have discussed this behavior in terms of a hopping-type relaxation in localized tail states. It is not easy to distinguish which mechanism dominates the THz conductivity: dielectric relaxation or hopping. If τ_{min} changes with temperature or optical pumping intensity, the dynamic loss may originate from dielectric relaxation. There is unfortunately no literature available on this subject.

Photoluminescence

We briefly review the photoluminescence in a-Si:H and amorphous chalcogenides. A radiative photon emission process in condensed matter is called luminescence, and conceptually photoluminescence (PL) is the inverse process of the optical absorption already discussed. As all information on photon absorption and emission in materials can be involved in PL processes, PL is an important topic, and there are many reports and reviews discussing PL mechanisms in amorphous semiconductors (see, for example, Aoki, 2006a, 2006b, 2012).

Frequency-resolved spectroscopy (FRS) may be used to understand directly the lifetime distribution. In particular, *quadrature frequency-resolved spectroscopy* (QFRS) is known to be more suitable for studying amorphous semiconductors having a broad PL lifetime distribution (Aoki, 2006b). As shown

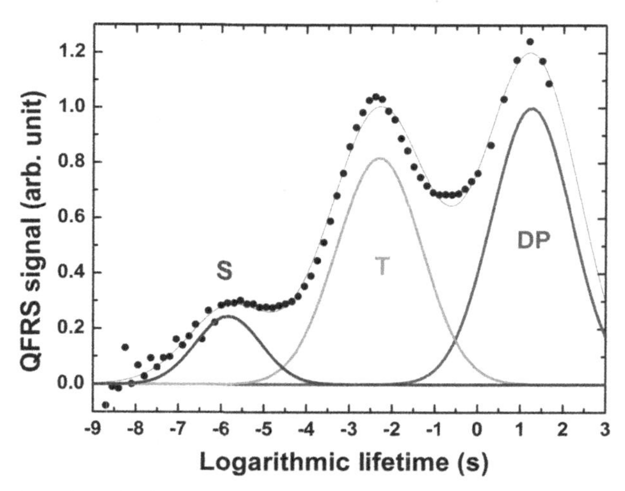

Figure 4.16. A broad photoluminescence lifetime distribution of a-Si:H. (S): singlet excitonic recombination; (T): triplet excitonic recombination; (DP): distant-pair recombination. (From Aoki (2006b). Photoluminescence. In *Optical Properties of Condensed Matter and Applications*, ed. J. Singh. Chichester: John Wiley & Sons, pp. 75–106. Copyright 2013 by John Wiley and Sons Inc.)

in Figure 4.16, Aoki and co-workers (Aoki, 2006a) succeeded in using QFRS to expand the very wide time (frequency) range 10^{-9}–10^{2} s in a-Si:H. There are three main peaks in the lifetime distribution. The shortest lifetime (around 10^{-6} s) and the medium lifetime (around 10^{-3} s) can be identified as singlet (S) and triplet (T) excitonic recombination. The longest lifetime (around 10^{1} s) can be explained by distant-pair recombination, in which photoexcited electrons and holes are randomly distributed in band tails.

Similar S and T peaks have been also found in a-Chs (Aoki *et al.*, 2005), and the principal features in PL have been universally observed in amorphous semiconductors.

Electronic transport (in dark conditions)

In this section, we discuss the electronic transport phenomena in a-Si:H and a-Chs. The electronic transport around room temperature is known to be dominated by band transport for both material systems. The central issues concerning the electronic transport have been widely discussed in the literature, and we do not need to consider them here because there is nothing new to report to date. However, we should say that there is an important and interesting issue regarding electronic transport that is still not properly understood. This is called the

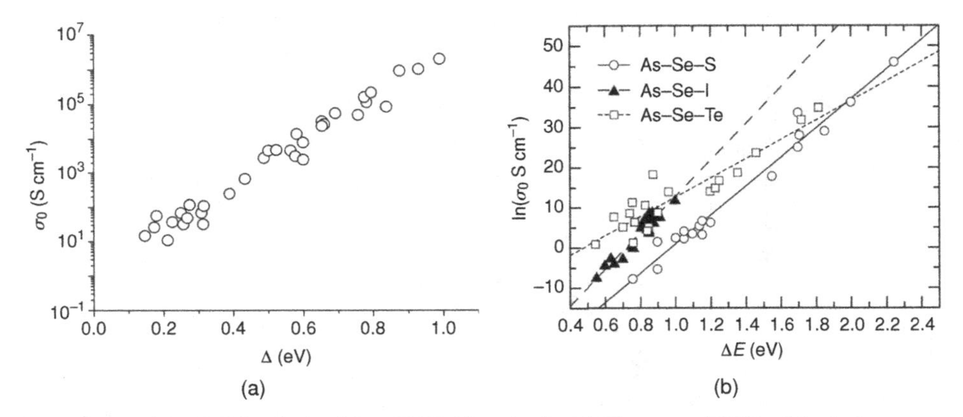

Figure 4.17. MN rule in (a) a-Si:H (Overhof and Thomas, 1989), (b) chalcogenide glasses (Shimakawa and Abdel-Wahab, 1997).

Meyer–Neldel (MN) rule or the compensation rule. A comprehensive review is given by Yelon, Movaghar, and Crandall (2006), and the current understanding is reported in Linert and Yelon (2013).

We therefore briefly introduce the MN rule as reported in electronic transport for a-Si:H and a-Chs, although the MN rule is related to chemical reactions, the dynamics of electronic and ionic transports, and thermodynamics in many material systems, when these mechanisms are thermally activated.

The band-like electronic conductivity, for example in a-Si:H and a-Chs, near room temperature may be empirically described as

$$\sigma = \sigma_{00} \exp(\Delta E / E_{MN}) \exp(-\Delta E / k_B T) \equiv \sigma_0 \exp(-\Delta E / k_B T), \quad (4.16)$$

where the prefactor σ_0 is not a constant; it is given by $\sigma_{00}\exp(\Delta E/E_{MN})$. Here, ΔE and E_{MN} are called the activation energy and the Meyer–Neldel energy, respectively. Examples of the MN rule are shown in Figures 4.17(a) and (b) for a-Si:H and a-Chs, respectively. The prefactor itself is a function of the activation energy ΔE. The MN energy, E_{MN}, drops around 40–50 meV.

In a-Si:H, this behavior is well explained by the statistical shift of the Fermi level (i.e. the temperature variation of the Fermi level) (Overhof and Thomas, 1989). Although, as shown in Figure 4.17(b), a similar effect can be found in the a-Chs, this behavior is not easy to explain using the statistical shift of the Fermi level (Shimakawa and Abdel-Wahab, 1997). As the MN rule is *universally* observed in a wide class of materials, Yelon *et al.* (2006) have suggested that the multiexcitation entropy is important in some kinetics and thermodynamics.

A recent idea proposed by Okamoto *et al.* (2010) is based on the universality of these phenomena.

In our view, however, *kinetics* and *thermodynamics* are different topics, and therefore each phenomenon may have different origins. For example, for ionic transport in crystalline solids, the same MN rule as in eqn. (4.16) has been found, and this can be explained in terms of phonon absorption and emission processes with lattice distortion (Shimakawa and Aniya, 2013). Note that the MN law is observed in both crystalline and disordered materials, although it is believed to exist only in disordered matter. The definitive solution to the important and unresolved problem of the MN rule is therefore extremely necessary, since this effect may contain unknown principles of physics.

References

Abkowitz, M. and Markovics, J.M. (1984). Evidence of equilibrium native defect populations in amorphous chalcogenides from analyses of xerographic spectra. *Philos. Mag. B*, **49**, L31–L36.

Anderson, P.W. (1958). Absence of diffusion in certain random lattices. *Phys. Rev.*, **109**, 1492–1505.

Anderson, P.W. (1975). Model for the electronic structure of amorphous semiconductors. *Phys. Rev. Lett.*, **34**, 953–955.

Aoki, T. (2006a). Understanding the photoluminescence over 13-decade lifetime distribution in a-Si:H. *J. Non-Cryst. Solids*, **352**, 1138–1143.

Aoki, T. (2006b). Photoluminescence. In *Optical Properties of Condensed Matter and Applications*, ed. J. Singh. Chichester: John Wiley & Sons, p. 75.

Aoki, T. (2012). Photoluminescence spectroscopy. In *Characterization of Materials*, 2nd edn., ed. E.N. Kaufmann. Hoboken, NJ: John Wiley & Sons, pp. 1–12.

Aoki, T., Saito, D., Ikeda, K., Kobayashi, S., and Shimakawa, K. (2005). Radiative recombination processes in chalcogenide glasses deduced by lifetime measurements over 11 decades. *J. Opt. Adv. Mater.*, **7**, 1749–1757.

Bishop, S.G., Strom, U., and Taylor, P.C. (1975). Optically induced localized paramagnetic states in chalcogenide glasses. *Phys. Rev. Lett.*, **34**, 1346–1350.

Chittik, R.C., Alexander, J.H., and Sterling, H.J. (1969). The preparation and properties of amorphous silicon. *J. Electrochem. Soc.*, **116**, 77–81.

Dyre, J.C. (1988). The random free energy barrier model for ac conduction in disordered solids. *J. Appl. Phys.*, **64**, 2456–2468.

Dyre, J.C. and Schroder, T.B. (2000). Universality of ac conduction disordered solids. *Rev. Mod. Phys.*, **72**, 873–892.

Fekete, L., Kuzel, P., Nemec, H. *et al.* (2009). Ultrafast carrier dynamics in microcrystalline silicon probed by time-resolved terahertz spectroscopy. *Phys. Rev. B*, **79**, 115306, 1–13.

Ganjoo, A. and Shimakawa, K. (1994). Estimation of density of charged defects in amorphous chalcogenides from a.c. conductivity: random-walk approach for bipolarons based on correlated barrier hopping. *Philos. Mag. Lett.*, **70**, 287–291.

He, X.-F. (1990). Fractional dimensionality and fractional derivative spectra of interband optical transitions. *Phys. Rev. B*, **42**, 11751–11756.

Hirabayashi, I. and Morigaki, K. (1983). Light induced metastable effect on the short lived photoinduced midgap absorption in hydrogenated amorphous silicon. *J. Non-Cryst. Solids*, **59** & **60**, 433–436.

Imagawa, O., Akiyama, T., and Shimakawa, K. (1984). Localized density of states in amorphous silicon determined by electrophotography. *Appl. Phys. Lett.*, **45**, 438–439.

Isoya, J., Yamasaki, S., Ohkushi, H., Matsuda, A., and Tanaka, K. (1993). Electron-spin-echo envelope-modulation study of the distance between dangling bonds and hydrogen atoms in hydrogenated amorphous silicon. *Phys. Rev. B*, **47**, 7013–7024.

Jackson, W.B. and Amer, N.M. (1982). Direct measurement of gap-state absorption in hydrogenated amorphous silicon by photothermal deflection spectroscopy. *Phys. Rev. B*, **25**, 5559–5562.

Kádas, K. and Kugler, S. (1997a). Midgap states in nitrogen doped diamond-like amorphous carbon. *Solid State Commun.*, **102**, 721–723.

Kádas, K. and Kugler, S. (1997b). Impurity levels in phosphorus and boron doped amorphous silicon. *Phil. Mag. B*, **76**, 281–285.

Kádas, K., Ferenczy, Gy.G., and Kugler, S. (1998). Theory of dopant pairs in four-fold coordinated amorphous semiconductors. *J. Non-Cryst. Solids*, **227–230**, 367–371.

Kastner, M., Adler, D., and Fritzsche, H. (1976). Valence-alternation model for localized gap states in lone-pair semiconductors. *Phys. Rev. Lett.*, **37**, 1504–1507.

Kugler, S. (2012). Advances in understanding the defects contributing to the tail states in pure amorphous silicon. *J. Non-Cryst. Solids*, **358**, 2060–2062.

Kugler, S. and Laszlo, I. (1989). Connection between topology and π-electron structure in amorphous carbon. *Phys. Rev. B*, **39**, 3882–3884.

Kugler, S., Kohary, K., Kádas, K., and Pusztai, L. (2003). Unusual atomic arrangements in amorphous silicon. *Solid State Commun.*, **127**, 305–309.

Lang, D.V. (1974). Deep-level transient spectroscopy: a new method to characterize traps in semiconductors. *J. Appl. Phys.*, **45**, 3023–3032.

Laszlo, I. (1993). Stable electronic energy levels in the presence of off-diagonal disorder. *Int. J. Quantum Chem.*, **48**, 135–146.

Laszlo, I. (2000). Graph theoretical study of topologically determined electronic energy levels. *J. Mol. Struct.: THEOCHEM*, **501–502**, 501–508.

Linert, W. and Yelon, A. (eds.) (2013). Isokinetic relationships. *Monatsh. Chem.* (special issue), **144**, 1–2.

Lloyd-Hughes, J. and Tae-In Jeon (2012). A review of the terahertz conductivity of bulk and nano-materials. *J. Infrared Milli. Terahz. Waves*, **33**, 871–926.

Lukács, R., Hegedus, J., and Kugler, S. (2008). Structure and photoinduced volume changes of obliquely deposited amorphous selenium. *J. Appl. Phys.*, **104**, 103512, 1–6.

Lukács, R., Hegedus, J., and Kugler, S. (2009). Microscopic and macroscopic models of photo-induced volume changes in amorphous selenium. *J. Mater. Sci: Mater. Electron.*, **20**, S33–S37.

Mackenzie, K.D., LeComber, L.G., and Spear, W.E. (1982). The density of states in amorphous silicon determined by space-charge-limited current measurements. *Philos. Mag. B*, **46**, 377–389.

Madan, A., LeComber, P.G., and Spear, W.E. (1976). Investigation of the density of localized states in a-Si using the field effect technique. *J. Non-Cryst. Solids*, **20**, 239–257.

Mandelbrot, B.B. (1982). *The Fractal Geometry of Nature*. New York: Freeman.

Marshall, J.M. and Owen, A.E. (1976). Field-effect measurements in disordered $As_{30}Te_{48}Si_{12}Ge_{10}$ and As_2Te_3. *Philos. Mag.*, **33**, 457–474.

Morigaki, K. (1999). *Physics of Amorphous Semiconductors*. London: Imperial College Press and World Scientific.

Mott, N.F. and Davis, E.A. (1979). *Electronic Processes in Non-Crystalline Materials*, 2nd edn. Oxford: Clarendon Press, p. 272.

Nessa, M., Shimakawa, K., Ganjoo, A., and Singh, J. (2000). Fundamental optical absorption on fractals: a case example for amorphous chalcogenides. *J. Optoelectron. Adv. Mater.*, **2**, 133–138.

Okamoto, H., Sobajima, Y., Toyama, T., and Matsuda, A. (2010). Laplace Meyer–Neldel relation. *Phys. Status Solidi A*, **207**, 566–569.

Overhof, H. and Thomas, P. (1989), *Electronic Transport in Hydrogenated Amorphous Semiconductors*. Berlin: Springer-Verlag.

Pantelides, S.T. (1986). Defects in amorphous silicon: a new perspective. *Phys. Rev. Lett.*, **57**, 2979–2982.

Petri, I., Salmon, P.S., and Fischer, H.E. (2000). Defects in a disordered world: the structure of glassy GeSe2. *Phys. Rev. Lett.*, **84**, 2413–2416.

Rayment, T. and Elliott, S.R. (1983). Small-angle neutron scattering study of anisotropic growth morphology and irreversible photodensification in *a*-$GeSe_3$ films. *Phys. Rev. B*, **28**, 1174–1177.

Shimakawa, K. and Abdel-Wahab, F. (1997). The Meyer–Neldel rule in chalcogenide glasses. *Appl. Phys. Lett.*, **70**, 652–654.

Shimakawa, K. and Aniya, M. (2013). Dynamics of atomic diffusion in condensed matter: origin of the Meyer–Neldel compensation law. *Monatsh. Chem.*, **144**, 67–71.

Shimakawa, K. and Ganjoo, A. (2002). ac photoconductivity of hydrogenated amorphous silicon: influence of long-range potential fluctuations. *Phys. Rev. B*, **65**, 165213, 1–5.

Shimakawa, K. and Katsuma, Y. (1986). Extended step-by-step analysis in space-charge-limited current: application to hydrogenated amorphous silicon. *J. Appl. Phys.*, **60**, 1417–1421.

Shimakawa, K., Kolobov, A.V., and Elliott, S.R. (1995). Photoinduced effects and metastability in amorphous semiconductors and insulators. *Adv. Phys.*, **44**, 475–588.

Shimakawa, K., Singh, J., and O'Leary, S.K. (2006). Optical properties of disordered condensed matter. In *Optical Properties of Condensed Matter and Applications*, ed. J. Singh. Chichester: John Wiley & Sons, pp. 47–62.

Shimakawa, K., Wagner, T., and Frumar, M. (2013). THz photoconductivity in a-Si:H. *Phys. Status Solidi B*, **250**, 1004–1007.

Singh, J. (2002). Effective mass of charge carriers in amorphous semiconductors and its applications. *J. Non-Cryst. Solids*, **299–302**, 444–448.

Singh, J. and Shimakawa, K. (2003). *Advances in Amorphous Semiconductors*. London: Taylor & Francis, p. 57.

Sólyom, J. (2009). *Fundamentals of the Physics of Solids. Volume 2: Electronic Properties*. Berlin Heidelberg: Springer-Verlag, p. 220.

Spear, W.E. and Le Comber, P.G. (1975). Substitutional doping of amorphous silicon. *Solid State Commun.*, **17**, 1193–1196.

Starbova, K., Mankov, V., Dikova, J., and Starbov, N. (1999). The effects of vapour incidence on the microstructure and related properties of condensed GeS_2 thin films. *Vacuum*, **53**, 441–446.

Street, R.A. (1991). *Hydrogenated Amorphous Silicon*. Cambridge: Cambridge University Press.

Tanaka, K. and Shimakawa, K. (2011). *Amorphous Chalcogenide Semiconductors and Related Materials*. New York: Springer.

Vanecek, M., Kocka, J., Stuchlik, J., Kozisek, Z., Stika, O., and Triska, A. (1983). Density of the gap states in undoped and doped glow discharge a-Si:H. *Solar Energy Mater.*, **8**, 411–423.

Weaire, D. and Thorpe, M.F. (1971). Electronic properties of an amorphous solid. I. A simple tight-binding theory. *Phys. Rev. B*, **4**, 2508–2520.

Yelon, A., Movaghar, B., and Crandall, R.S. (2006). Multi-excitation entropy: its role in thermodynamics and kinetics. *Rep. Prog. Phys.*, **69**, 1145–1194.

Zallen, R. (1983). *The Physics of Amorphous Solids*. New York: John Wiley & Sons Inc.

5

Photoinduced phenomena

Reversible and *irreversible* photoinduced changes are found in amorphous semiconductors, particularly in amorphous chalcogenides (a-Chs). The term *reversible* usually means that the final state of a substance corresponds to its initial state by means of structural relaxation. The structural relaxation is triggered by switching off the illumination, thermal annealing near the glass-transition temperature, or other photon irradiation. Reversible photoinduced phenomena can be seen as the changes in the physical properties of the semiconductor. On the other hand, irreversible changes are mostly related to chemical reactions or to a phase change, such as crystallization. We are interested in physical changes in a-Chs, and will mostly concentrate on the reversible phenomena observed in amorphous semiconductors.

5.1 Photoinduced volume changes

In amorphous chalcogenides, photodarkening (PD, in which the bandgap decreases during and after the illumination) and photoinduced volume changes (usually photoinduced volume expansion, PVE) were for a long time believed to be two aspects of the same phenomenon. Therefore, a one-to-one correlation was expected to exist between PD and PVE. The first systematic study of PVE was performed by Hamanaka *et al.* (1976), showing that PD is always accompanied by PVE. However, the experiment carried out by Tanaka (1998) showed that from the time evolution of these two phenomena it follows that the time constants of the PD and the PVE in a-As_2S_3 are different. It was observed that the evolution of PVE saturates earlier than that of PD, suggesting that these two phenomena are not directly related to each other.

As most of the measurements of *reversible* PVE have been made after switching off the illumination (i.e. in the metastable state) (Hamanaka *et al.*, 1976; Tanaka, 1990, 1998; Kuzukawa *et al.*, 1999), the correlation during illumination remains unclear. To understand the dynamics of PVE and PD *during* illumination, *in situ* measurements were performed (Ganjoo, Ikeda, and Shimakawa, 1999; Ganjoo *et al.*, 2000: Ganjoo and Shimakawa, 2002). A real-time *in situ* surface height measuring system has been developed using a Twyman–Green interferometer with image analysis technology (Ikeda and Shimakawa, 2004; Shimakawa, Ikeda, and Kugler, 2004a).

Figure 5.1(a) shows an example of a surface height map for well-annealed flatly deposited a-As_2Se_3 films (on a Si substrate). The measurements were carried out at 300 K. Figures 5.1(b) and (c) show the time evolution of the changes in flatly and obliquely deposited films. In flatly deposited films (Figure 5.1(b)), the surface height increased by 10 nm ($\Delta d/d = 2\%$) during 200 s of laser illumination (wavelength of 532 nm and power density of 90 mW cm^{-2}). After 600 s, the illumination was turned off. The surface height started decreasing and settled in 200 s, at which point the height was 2 nm lower than before the light was switched off. In the case of obliquely deposited films (Figure 5.1(c)), the surface height decreased by 12 nm ($\Delta d/d = 2.4\%$) in 3×10^4 s. Similar results have been observed in other a-Chs.

As shown in Figure 5.2, a less significant PVE than that seen in a-As_2Se_3 was found in (flatly deposited) a-Se films. The height increased rapidly by 2.5 nm ($\Delta d/d = 0.5\%$) with the illumination (wavelength of 532 nm in wavelength and power density of 90 mW cm^{-2}). As soon as the light was turned off, after 800 s of illumination, the height decreased by 2 nm and then gradually decreased to the original height in 200 s.

Let us briefly discuss the height variations. As shown in Figure 5.1, in flatly deposited a-As_2Se_3 films PVE was observed *during and after* the illumination. The transient PVE (PVE occurs only during the illumination) must be involved in the whole PVE process (Ganjoo *et al.*, 2000). The remaining increase in surface height that occurred after illumination is a so-called "metastable" PVE (Shimakawa, Kolobov, and Elliott, 1995; Tanaka, 1998). In obliquely and as-deposited a-As_2Se_3 films, as shown in Figure 5.1(c), the surface decreased with time when the light was turned on (volume contraction). The PVE may have been taken over by photoinduced volume contraction (PVC), and hence the height decreased with time. This volume contraction can be interpreted by considering void corruption, as many voids are involved in obliquely

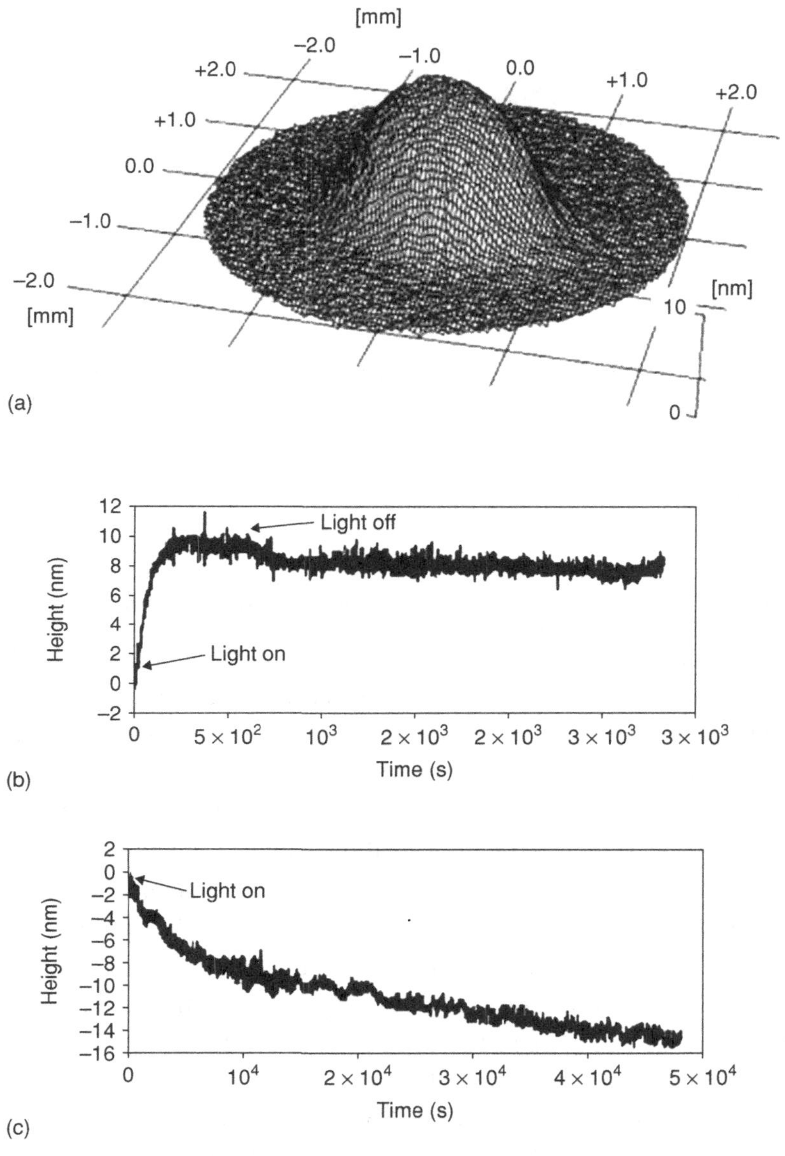

Figure 5.1. (a) Surface height map for flatly deposited a-As_2Se_3. (b) Time evolution of surface height for flatly deposited a-As_2Se_3. (c) Time evolution of surface height for obliquely deposited a-As_2Se_3. (Adapted from Ikeda and Shimakawa (2004). *J. Non-Cryst. Solids*, **338–340**, 539. Copyright 2013 with permission from Elsevier.)

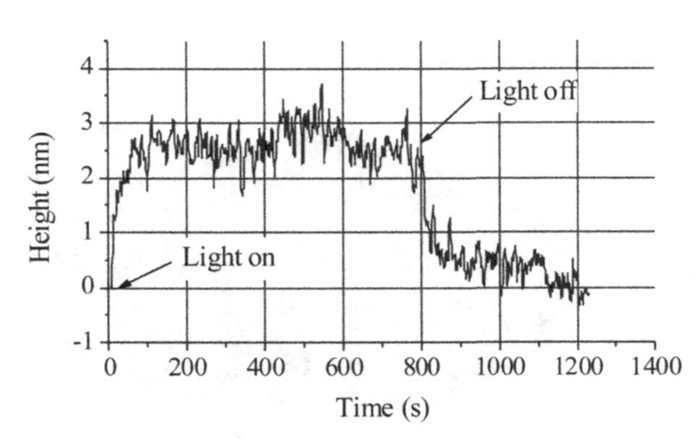

Figure 5.2. Time evolution of surface height for flatly deposited a-Se.

deposited films which consist of a columnar structure with a large amount of free space.

Atomic-scale computer simulations of photoinduced volume changes

The photoinduced volume changes in a-Se were simulated using the TBMD computer code ATOMDEP (see Chapter 3). A tight-binding model developed by Molina, Lomba, and Kahl (1999) was used to describe the atomic interactions between selenium atoms. The velocity Verlet algorithm was applied in order to follow the motion of the atoms with a time step $t = 2$ fs. The temperature was controlled via the velocity-rescaling method. Samples containing 162 atoms were prepared by the "cook and quench" procedure. A temperature of 20 K was reached at the end of the preparation, and the sample for 500 MD steps (1 ps) was relaxed. In order to model the photoinduced volume changes, the periodic boundary conditions were lifted along the z direction at this point. This procedure gave us a slab geometry with periodic boundary conditions in two dimensions only. The system was then relaxed for another 40 000 MD steps (80 ps) to obtain a stable configuration. The sample can expand or shrink in the z direction. As a measure of sample thickness, the distance between the centers of masses of 15–15 surface atoms on both sides was used, as indicated by the the arrow in Figure 5.3 (Lukács, Hegedus, and Kugler, 2009).

Photoexcitation generates an electron–hole pair when a photon is absorbed. This process can be modeled by transferring an electron from the highest occupied molecular orbital (HOMO) to the lowest unoccupied molecular orbital (LUMO). In the computer simulations it was assumed that, immediately after

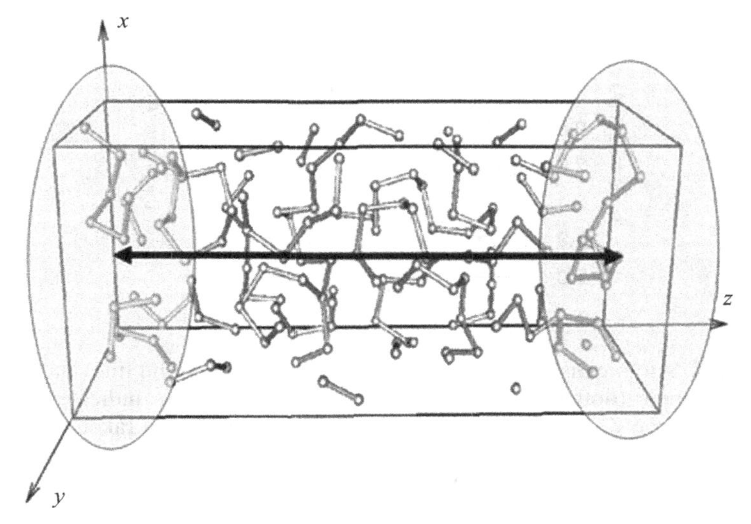

Figure 5.3. Snapshot of a glassy selenium network. The sample can move in the z direction. (Adapted from Lukács, Hegedus, and Kugler (2009). *J. Mater. Sci.: Mater. Electron.*, **20**, S33. With kind permission from Springer Science and Business Media.)

the photon absorption, the electron and the hole became separated in space on a femtosecond time scale (Hegedus *et al.*, 2005). The Coulomb interaction between the electron and the hole was neglected. The influences due to the excited electron and the hole were treated independently. At first, an extra electron was put into the LUMO (excited electron creation), then an electron in the HOMO (hole creation) was annihilated. When an additional electron was put in the LUMO, a bond-breaking event occurred. In the majority of cases, a covalent bond between twofold- and threefold-coordinated atoms was broken; however, bond breaking was sometimes observed between twofold- and twofold-coordinated atoms. Our localization analysis revealed that the LUMOs are usually localized at such sites before the bond breaking. The change in bond length alternates between shrinkage and elongation in the vicinity of the broken bond due to bond breaking. If the electron in the LUMO is de-excited, then all the bond lengths are restored to their original value. A characteristic time development of a single photoinduced bond-breaking event is shown in Figure 5.4. Before the excitation at 5 ps, the bond length was approximately 0.255 nm. During illumination, this bond length increased by 10–20%, then decreased to its original value after the de-excitation of the electron at 15 ps. The volume change of the a-Se slab followed the bond breaking, and it showed damped oscillations

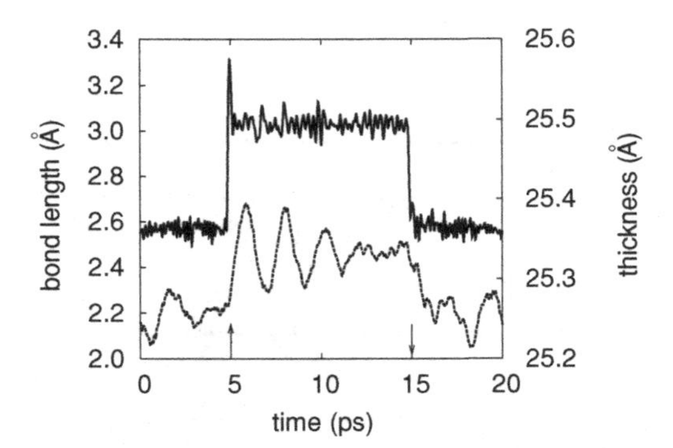

Figure 5.4. Atomic distance separation of breaking bond (solid line) and thickness of sample (dotted line) as a function of time. Arrows indicate that an excited electron was created at 5 ps and annihilated at 15 ps. (Taken with permission from Hegedus *et al.* (2005). *Phys. Rev. Lett.*, **95**, 206803. Copyright 2013 by the American Physical Society.)

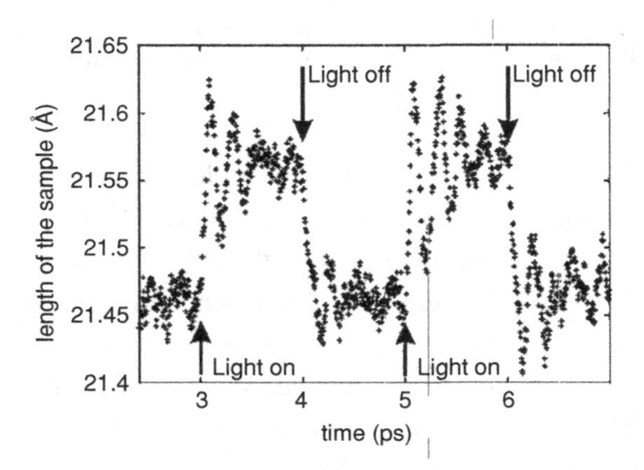

Figure 5.5. Reversible thickness change of a-Se sample during photoexcitation. (Adapted from Kugler, Hegedus, and Kohary (2007). *J. Mater. Sci.: Mater. Electron.*, **18**, S163. With kind permission from Springer Science and Business Media.)

on a picosecond time scale, as shown in Figure 5.4. The reversible thickness change of the a-Se sample was observed during several photoexcitations, as demonstrated in Figure 5.5 (Kugler, Hegedus, and Kohary, 2007).

Hartree–Fock *ab initio* Raman spectra calculations and Raman spectroscopic measurements were carried out on a-Se to identify the characteristic vibrational mode caused by sigma bonds (Lukács *et al.*, 2010). In the Raman spectra the peaks are around 250 cm^{-1} (corresponding to 0.234 nm), which is attributed

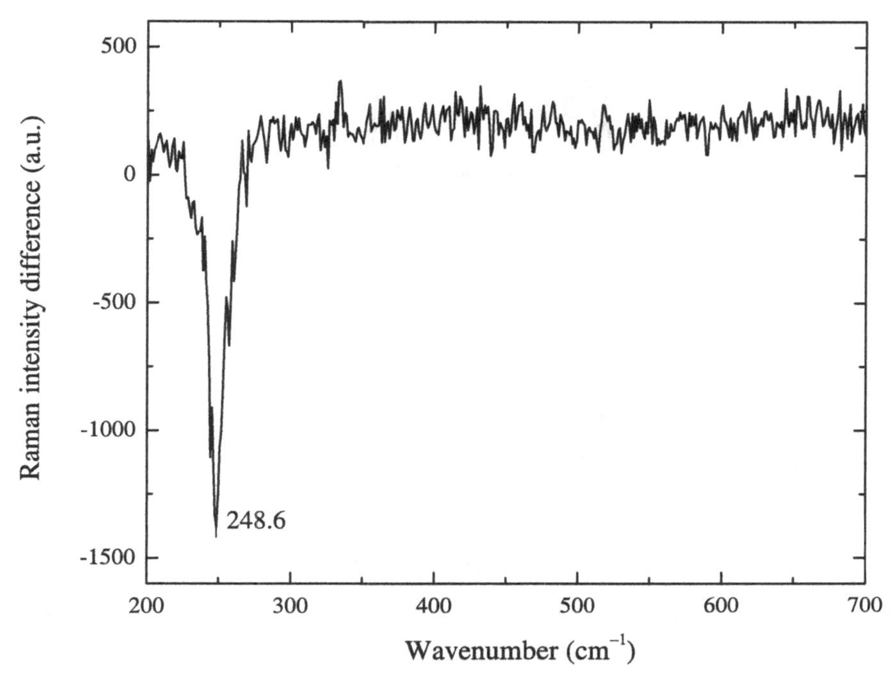

Figure 5.6. Difference between the Raman spectra of the 10 min illuminated a-Se sample and that of the 40 min relaxed a-Se sample. (Taken with permission from Lukács *et al.* (2010). *J. Appl. Phys.*, **107**, 073517. Copyright 2013, American Institute of Physics.)

to the vibrational modes of the covalent bonds in a-Se. As the Raman intensity varies in time due to the illumination, a large number of covalent bonds break; these form again once the illumination is switched off. The difference between the Raman spectra of the sample illuminated for 10 min and that for the relaxed sample after 40 min is shown in Figure 5.6. The Raman spectra measurement provides strong experimental evidence for the predicted photoinduced bond-breaking process.

A very different behavior is observed during hole creation (Hegedus *et al.*, 2005). Interchain weak bonds (see Figure 3.23) are formed after creating a hole, thereby causing the contraction of the sample, as observed in Figure 5.7. This always occurs near the atoms where the HOMO is localized. Since the HOMO is usually localized in the vicinity of a onefold-coordinated atom, the interchain bond formation often takes place between a onefold-coordinated atom and a twofold-coordinated atom. Sometimes, the formation of interchain bonds between two twofold-coordinated atoms is also observed.

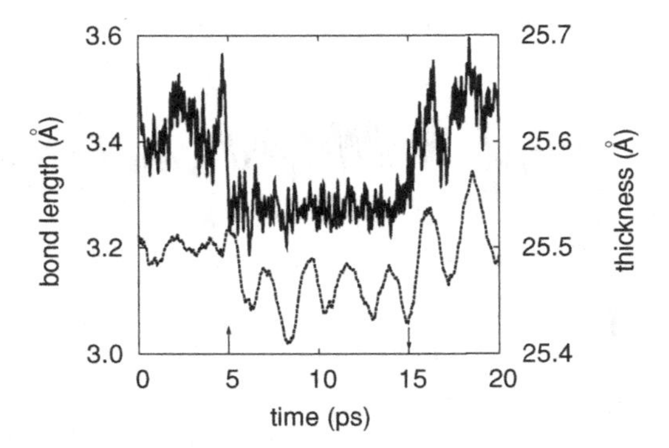

Figure 5.7. Atomic distance separation of interchain bond (solid line) and thickness of sample (dotted line) as a function of time. (Taken with permission from Hegedus *et al.* (2005). *Phys. Rev. Lett.*, **95**, 206803. Copyright 2013 by the American Physical Society.)

Kinetics of volume change

Volume expansion and shrinkage are additive quantities, i.e. the expansion in thickness d_+ is proportional to the number of excited electrons n_e and is given by $d_+ = \beta_+ n_e$. Similarly, the shrinkage d_- is proportional to the number of created holes n_h and is given by $d_- = \beta_- n_h$. The parameter β_+ (β_-) is the average thickness change caused by an excited electron (hole). The time-dependent thickness change is equal to $d(t) = d_+(t) - d_-(t) = \beta_+ n_e(t) - \beta_- n_h(t)$. Assuming that the number of electrons is equal to the number of holes, $n_e(t) = n_h(t) = n(t)$, we obtain

$$d(t) = (\beta_+ - \beta_-)n(t) = \beta n(t), \tag{5.1}$$

where β is a characteristic constant of the chalcogenide glass related to photoinduced volume change, and is a unique parameter for each sample. The sign of this parameter determines whether the material expands or shrinks.

The number of photoexcited electrons and holes is proportional to the time of a steady state illumination. Their generation rate G depends on the photon absorption coefficient and the number of incoming photons. After photon absorption, the excited electrons and holes randomly diffuse and eventually recombine. A phenomenological non-linear rate equation for this process is given by

$$\frac{dn\,(t)}{dt} = G - Cn(t)^2. \tag{5.2}$$

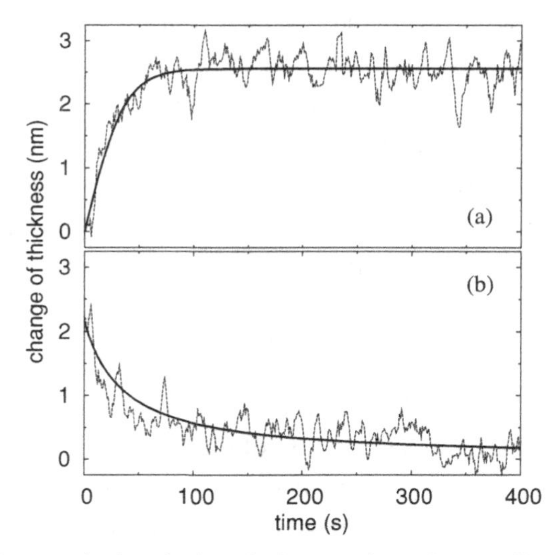

Figure 5.8. Measured photoinduced changes in a-Se. (a) Expansion due to illumination (dotted line) and theoretical (tanh) curve (solid line). (b) Shrinkage after switching off illumination (dotted line) and fitted theoretical curve (solid line). (Taken with permission from Hegedus *et al.* (2005). *Phys. Rev. Lett.*, **95**, 206803. Copyright 2013 by the American Physical Society.)

The solution for the time-dependent volume change is given by

$$d(t) = \beta\sqrt{\frac{G}{C}}\tanh\left(\sqrt{GC}\,t\right). \tag{5.3}$$

The factor $\tanh(x)$ is measurable after a lengthy illumination ($\tanh(\infty) = 1$). Only one fitting parameter remains ($\sqrt{GC}$). After the light has been turned off, the rate equation reduces to

$$\frac{dn(t)}{dt} = -Cn(t)^2, \tag{5.4}$$

which has the solution

$$d(t) = \frac{a}{(aCt/\beta) + 1}, \tag{5.5}$$

where a is equal to $d(t)$ when the illumination is switched off. The photoinduced expansion of a-Se films was measured *in situ* using optoelectronic interference, enhanced by image processing (Ikeda and Shimakawa, 2004). Figures 5.8(a) and (b) show the measured time evolution and decay, respectively, of the surface height; the best fits were obtained in the 0–400 s interval of illumination and after switching off the light.

Some glasses show irreversible changes during illumination (Hegedus, Kohary, and Kugler, 2006). Furthermore, the measured photoinduced volume changes in obliquely and flatly deposited a-AsSe were quite different; i.e. the first one shrunk and the other one expanded (Ikeda and Shimakawa, 2004). To explain this difference, we must take into account the large number of irreversible changes in the local atomic arrangement. The total expansion includes reversible (transient) and irreversible (metastable) changes:

$$d(t) = d_{rev}(t) + d_{irr}(t). \tag{5.6}$$

The reversible part follows the previously described time development during and after illumination. There is no volume change caused by irreversible microscopic effects after switching off the light. During the illumination, the irreversible part of expansion/shrinkage is governed by

$$\frac{dn(t)}{dt} = G(N_0 - n(t)), \tag{5.7}$$

where G describes the excited electron–hole generation rate and N_0 is the maximum number of sigma bonds due to the irreversible bond breaking. Rewriting the mathematical equation, we obtain

$$\frac{d(d_{irr}(t))}{dt} = G_{irr} - C_{irr}d_{irr}(t). \tag{5.8}$$

The solution is given by

$$d_{irr(t)} = \frac{G_{irr}}{C_{irr}}(1 - \exp(-C_{irr}t)). \tag{5.9}$$

The best fits of photoinduced volume changes during and after illumination for flatly deposited a-AsSe are displayed in Figure 5.9. Both expressions have positive prefactors; i.e. during the illumination, the sample expands.

The obliquely deposited porous sample shrinks. We can describe this process using the same mathematical expression, but now both expressions have negative prefactors (Lukács and Kugler, 2011). The best fit for obliquely deposited a-AsSe is shown in Figure 5.10. Negative prefactors mean that both reversible and irreversible bond breaking cause macroscopic shrinkage in porous materials (Chapter 3).

We should make an additional remark about time-dependent behavior. Instead of a Debye-type normal exponential relaxation, the stretched exponential

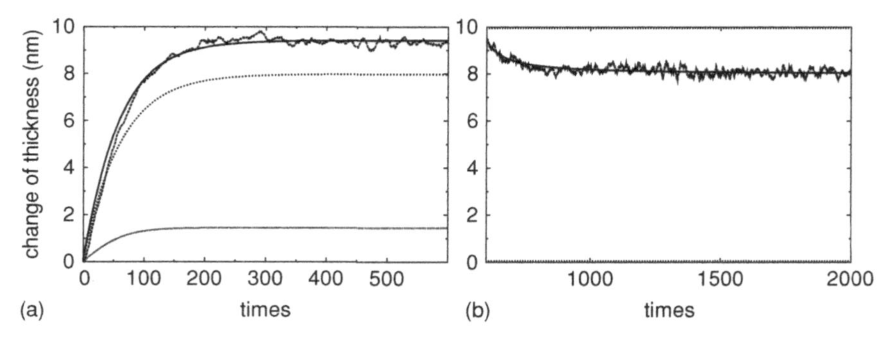

Figure 5.9. Volume changes as a function of time of flatly deposited a-AsSe thin film. Solid thin lines are the measured volume changes (a) during and (b) after illumination; the solid thick line is the best fit; the two dotted lines represent the reversible and irreversible parts of the theoretical curve. (From Hegedus, Kohary, and Kugler (2006). *J. Non-Cryst. Solids*, **352**, 1587. Copyright 2013 with permission from Elsevier.)

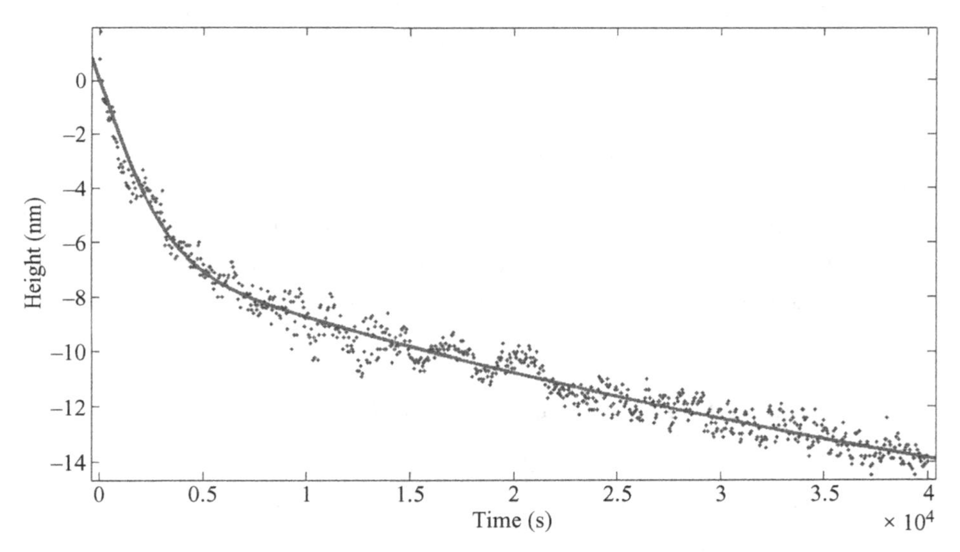

Figure 5.10. Volume changes as a function of time of obliquely deposited a-AsSe thin film. Symbols denote the measured volume changes during the illumination, and the solid line is the best fit. (From Lukács and Kugler (2011). *Jpn. J. Appl. Phys.*, **50**, 091401.)

functions in many cases are fitted to the measured data in disordered systems. The mathematical expression is given by

$$D(t) = A\left[1 - \exp\left\{-(t/\tau)^{\beta}\right\}\right], \tag{5.10}$$

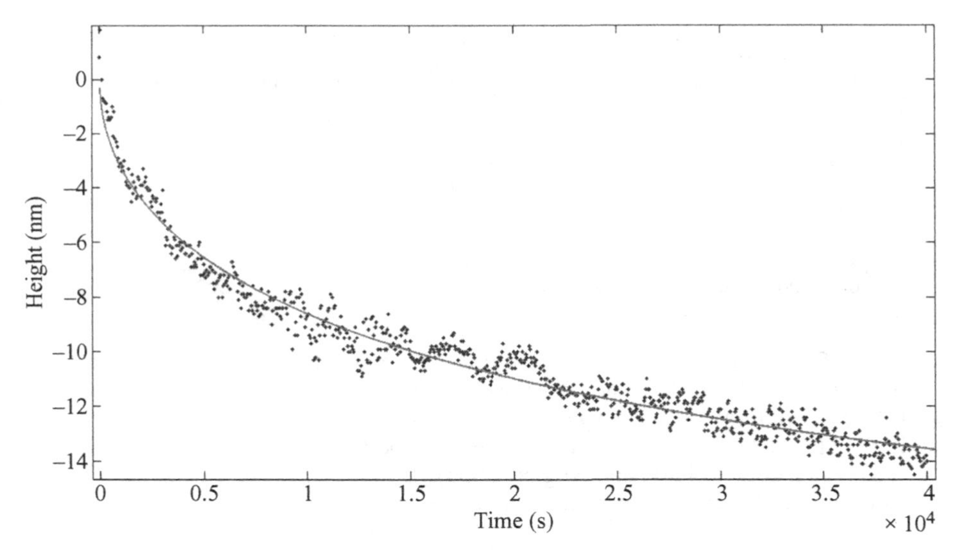

Figure 5.11. Volume changes of the same obliquely deposited a-AsSe thin film as in Figure 5.10 fitted by a stretched exponential function. (From Lukács and Kugler (2011). *Jpn. J. Appl. Phys.*, **50**, 091401.)

where β (<1), A, and τ are fitting parameters. Usually these curves fit quite well, yet the physical explanation of this phenomenon still remains unclear. In this case, the stretched exponential function also fits the exponential data quite well, as shown in Figure 5.11. This example demonstrates that two different processes (reversible and irreversible) can also be fitted by a stretched exponential function.

Photoinduced volume changes in a-Si:H films and oxide glasses

Finally, we discuss photoinduced volume changes observed in a-Si:H films and oxide glasses. Although a volume change after the bandgap illumination on a-Si:H is quite insignificant, PVE has been reported after employing a special technique, i.e. one that involves the detection of the bending of an optical lever, with a detection limit of $\Delta V/V = 2 \times 10^{-7}$ (Gotoh *et al.*, 1998). The time evolution of the PVE, together with the photoinduced defect creation (PDC) by an Ar-ion laser (300 mW cm^{-2}) is shown in Figure 5.12. The value of the PVE, $\Delta V/V$, approached 4×10^{-6}, which is much lower than the value of 4×10^{-3} found in a-As_2S_3. An intimate correlation between the dynamics of PVE and PDC (see Figure 5.12) suggests that PDC is the origin of PVE. The PDC in amorphous chalcogenides and a-Si:H will be discussed in Section 5.3.

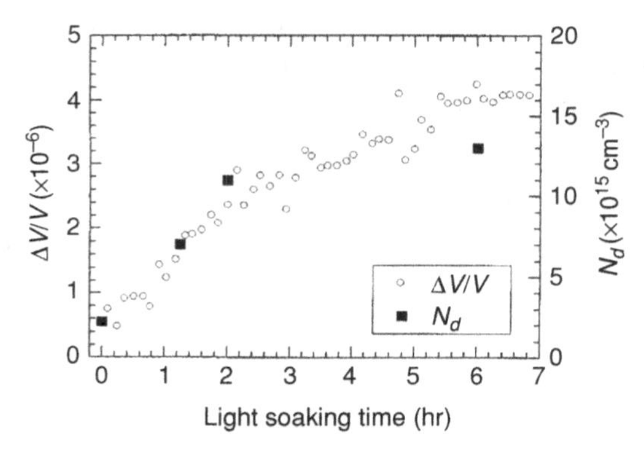

Figure 5.12. Time evolution of PVE together with PDC illuminated by Ar-ion laser (300 mW cm^{-2}) in a-Si:H. Open circles: volume change; closed squares: defect density. (Taken with permission from Gotoh *et al.* (1998). *Appl. Phys. Lett.*, **72**, 2978. Copyright 2013, American Institute of Physics.)

The opposite effect, i.e. volume compaction, occurs under bandgap illumination in g-SiO_2. Although the details are not clear, a possible mechanism is related to PDC (Uchino *et al.*, 2002). A photon creates a defect-like E$'$ center, which may trigger the volume contraction. It is of interest to compare this effect with volume expansion in a-Si:H.

5.2 Photodarkening and photobleaching

Photodarkening and photobleaching (PB) refer, respectively, to the bandgap decreasing and increasing with illumination. *Reversible* photodarkening (RPD) or photobleaching (RPB) is a characteristic feature of amorphous or glassy chalcogenide materials. These effects do not occur in crystalline chalcogenides. Note also that neither PD nor PB is observed in a-Si:H. As the refractive index n is known to be closely related to the bandgap of a material (Mott and Davis, 1979), the photoinduced changes in the refractive index fall into the category of PD or PB. Thus we do not discuss here the photoinduced optical constants n or k.

The PD in amorphous chalcogenides is a very popular subject and is discussed widely in numerous review works (see, for example, Tanaka, 1990; Pfeifer, Paesler, and Agarwal, 1991; Shimakawa *et al.*, 1995; Fritzshe, 2000; Kolobov, 2003). We start the discussion with an overall description of the features of the observed PD.

(1) Among many a-Chs, a-As_2S_3 shows the most significant change in the bandgap, E_g ($\Delta E_g \sim 50$ meV at around room temperature).
(2) ΔE_g decreases with increasing temperature and is scaled with T/T_g, where T is the temperature at which the illumination is made and T_g is the glass-transition temperature (Tanaka, 1983).
(3) Under the application of a high static pressure over 1.5 GPa (15 kbar), ΔE_g is known to decrease in a-Se (Tanaka, 1990).
(4) Bandgap illumination is the most effective for PD to occur.

In situ measurements are useful for examining PD dynamics (Ganjoo *et al.*, 2000). Therefore, we begin with the time evolution of the changes in the optical absorption coefficient $\Delta\alpha$. Two light beams were used: one was a light excitation (Ar laser) and the other one was a monitor beam (1.95 eV). The change in optical absorption coefficient is defined as $\Delta\alpha = (-1/d)\ln(\Delta T)$, where d is the film thickness and ΔT is the change in optical transmission, which is defined here as $\Delta T = T/T_0$ (where T is the transmission at any time t and T_0 is the transmission when the illumination is switched on). The details of the experimental set-up are described elsewhere (Ganjoo *et al.*, 2000).

Figure 5.13 is an example of the time evolution of $\Delta\alpha$ for a-As_2Se_3 films during various cycles of excitation at 50 and 300 K (Ganjoo *et al.*, 2002): $\Delta\alpha$ increased rapidly at first for both temperatures, before reaching a state close to saturation after some time. When the Ar-laser illumination was switched off, a decrease in $\Delta\alpha$ was observed, which quickly reached a constant value. This portion of the total change is the *transient* part induced by the illumination, and the portion which remained after stopping the illumination is the so-called "metastable" PD. The transient parts of the changes were found to be nearly 60% and 30% of the total changes induced during the illumination at 300 K and 50 K, respectively. The cycle was repeated many times after the metastable state was reached, and every illumination confirmed the occurrence of *transient* PD only.

5.3 Photoinduced defect creation: the Staebler–Wronski effect

Photoinduced degradation in a-Si:H was found during a decrease in photocurrent after illumination (Staebler and Wronski, 1977), which was then known as a phenomenon that accompanied defect creation (Street, 1991; Redfield and Bube, 1996; Morigaki, 1999). Subsequently, photoinduced defect creation has

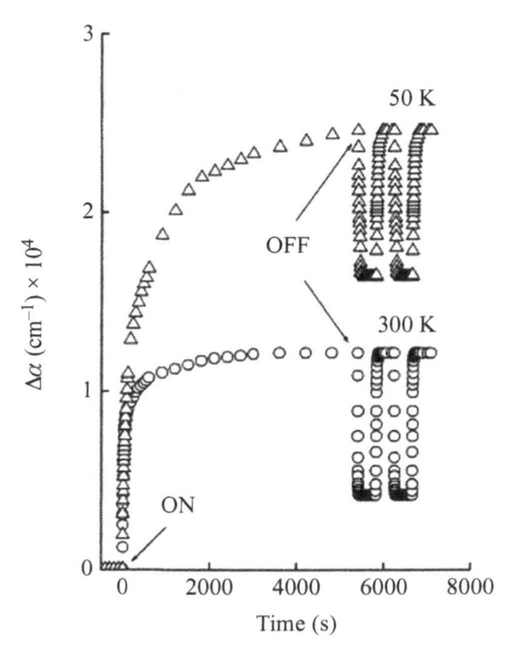

Figure 5.13. Time evolution of $\Delta\alpha$ for a-As_2Se_3 films during various cycles of excitation at 50 and 300 K. (From Ganjoo *et al.* (2002). *J. Non-Cryst. Solids*, **299–302**, 917. Copyright 2013 with permission from Elsevier.)

also been found in amorphous chalcogenides (Shimakawa, 1986; Shimakawa, Inami, and Elliott, 1990; Shimakawa *et al.*, 1992, 1995).

Electron spin resonance is a popular experimental technique that is used to understand the microscopic nature of defects, and therefore it applies to the study of the PDC process. It is observed that the number of ESR centers increases with illumination. However, a more convenient technique with which to obtain information concerning the change in the number of defects is photocurrent spectroscopy, in which a change in photocurrent (conductivity) produces a change in the number of defects (acting as recombination centers). As in a-Chs, the principal defects are not ESR active, so spectroscopy using photocurrent can be one of the more useful methods in the study of a-Chs as well as in a-Si:H (Shimakawa *et al.*, 1995).

First we discuss PDC in a-Si:H. The breaking of weak Si–Si bonds during illumination is a generally accepted model in which the breaking is induced by the non-radiative recombination of electrons and holes. A simple bond breaking produces a pair of neighboring dangling bonds (DBs). This model is not consistent with the ESR results, which evidently suggests the presence of well-separated DBs. It is now known that the photoenhanced diffusion of H atoms plays a role in PDC (Street, 1991; Branz, 1999; Morigaki, 1999; Morigaki and Hikita, 2007). The above models highlight the importance of hydrogen diffusion (H-diffusion) in a Si network which is assisted by photocarrier *recombination* energies. We should also consider the primary event of optical excitation; optical excitation itself may induce bond breaking. We do not have a proper answer to the question "Is a PDC process dominated by H-diffusion?". Consider, for example, the amorphous chalcogenides. Although the structural network is completely different from that of a-Si:H, PDC does occur in a-Chs. Therefore, it is suggested that the diffusion of hydrogen atoms is a secondary effect of PDC processes.

We now discuss PDC in a-Chs. Shimakawa *et al.* (1990, 1995) have assumed the existence of strong carrier–phonon interactions and have concluded that the primary event of photoexcitation is the creation of a self-trapped exciton (STE) and then its non-radiative recombination, leading either to the creation of metastable STE (or an intimate pair of charged defects) or to its return to the ground state through radiative recombination. This process is similar to that proposed in a-SiO_2 (Shugler and Stefanovich, 1990; Itoh and Stoneham, 2001). According to the STE model, the primary event is an optical excitation, not a non-radiative recombination. The weakening of a bond due to STE creation eventually leads to bond breaking. However, it should be noted that the creation of metastable STE is the result, not the cause, of bond breaking (Shugler and Stefanovich, 1990).

Let us discuss the model in detail. As already described in Section 4.3, it is widely accepted that defects in a-Chs are negative-U by their nature, and hence only the charged defects exist in this class of material (Mott and Davis, 1979). These defects are ESR inactive. However, optical excitation (a change in electron or hole occupation) changes a-Chs to ESR-active states. This is known as light-induced ESR (LESR).

A microscopic model for PDC, for example for a-$As_2S(Se)_3$, is shown in Figure 5.14. Optical excitation produces $P_2^+ - C_1^-$ and $P_4^+ - C_1^-$, which are called an intimate pair (IP) of defects, where P and C refer to pnictogen and chalcogen centers, respectively. The IPs created in this way can be regarded as the optically created STE (Shimakawa *et al.*, 1995; Song and Williams, 1996). The subsequent bond-switching reactions can lead to a significant separation between the charged defects, which are called random pairs (RPs). RPs are responsible for the decrease in the electronic properties (for example, photoconductivity,

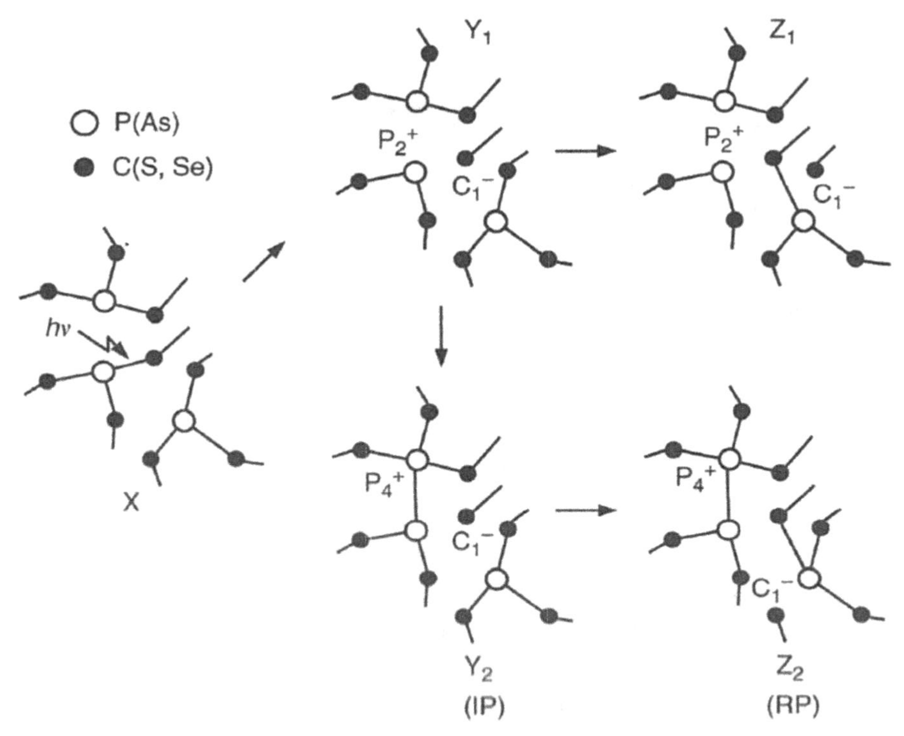

Figure 5.14. A microscopic model for PDC for a-$As_2S(Se)_3$. X is the ground state, Y_1 and Y_2 are the intimate pairs (IPs), and Z_1 and Z_2 are random pairs (RPs).

LESR, etc.) as well as the optical properties (Shimakawa *et al.*, 1995). Note that these IPs and RPs disappear after thermal annealing at low temperature (around 200 K) and at high temperature (near T_g). The details are described elsewhere (Singh and Shimakawa, 2003; Tanaka and Shimakawa, 2011).

Let us discuss the quantum efficiency of PDC. As already explained, the use of a photocurrent measurement during the illumination is useful in defect spectroscopy. Now let us focus on the defect creation $N(t)$ during the illumination. It is assumed that the change in the photocurrent is dominated by PDC itself. The photocurrent, under conditions of thermal equilibrium, together with the exposure time can be expressed as follows (Shimakawa *et al.*, 2004b):

$$I_p(t) = \frac{C}{N_0 + N(t)} = \frac{C}{N_0(1 + N(t)/N_0)} = \frac{I_p(0)}{1 + N(t)/N_0}, \qquad (5.11)$$

where C is a constant and $I_p(0)$ $(= C/N_0)$ is the initial photocurrent; $N(t)/N_0$ can be estimated from the ratio $I_p(0)/I_p(t)$, with N_0 being the initial number of defects.

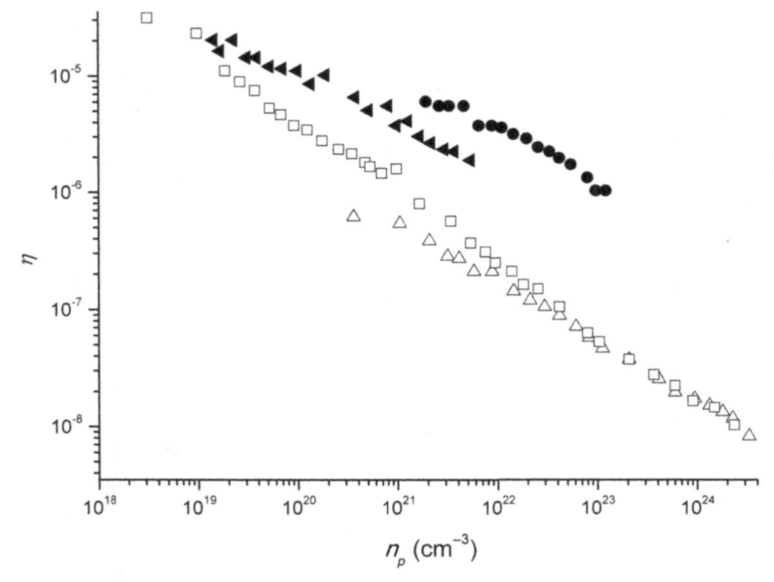

Figure 5.15. Quantum efficiency η of defect creation at 300 K for both a-As_2Se_3 (solid symbols) and a-Si:H (open symbols).

The photocurrent $I_p(t)$ is written empirically for both a-Chs and a-Si:H as follows (Shimakawa *et al.*, 1995, 2004b):

$$I_p(t) = \frac{I_s}{1 - \gamma \exp\left\{-(t/\tau)^{\beta}\right\}}, \tag{5.12}$$

where I_s is a constant current attained by prolonged illumination, $\gamma = N_s/(N_0 + N_s)$, τ is the effective creation time, and β is the dispersion parameter ($\beta < 1.0$). Here N_s is the saturated number of the defect density. A change in N_t is then given by

$$N(t) = N_s\left[1 - \exp\left\{-(t/\tau)^{\beta}\right\}\right]. \tag{5.13}$$

We now define the quantum efficiency (QE) η of PDC as

$$\eta = N(t)/Gt = N(t)/n_p, \tag{5.14}$$

where G is the rate of absorbed photons (cm^{-3} s^{-1}), t is the illumination time, and n_p (= Gt) is the total number of absorbed photons.

For comparison, Figure 5.15 shows the experimental data of η vs. n_p at 300 K for both a-As_2Se_3 (solid symbols) and a-Si:H (open symbols). The efficiency η decreased with increasing n_p and was higher for a-As_2Se_3 than for a-Si:H. This

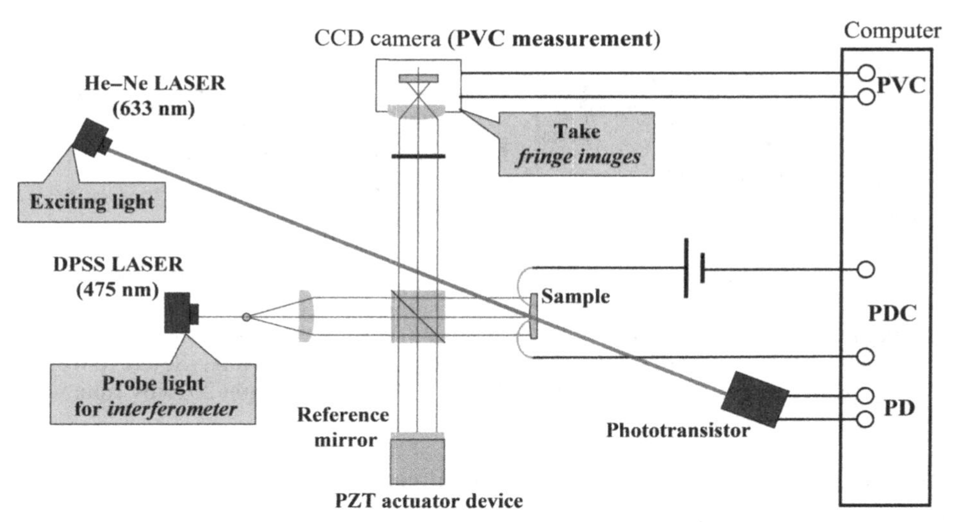

Figure 5.16. *In situ* simultaneous measurement system, which monitors the changes in surface height (for PVE), optical transmittance (for PD), and photocurrent (for PDC). (From Nakagawa *et al.* (2010). *Phys. Status Solidi C*, **7**, 857. Copyright 2013 by John Wiley and Sons Inc.)

can be attributed to the high degree of flexibility of the atomic structure in a-Chs. A reported QE value of $\eta \sim 10^{-8}$ is found for a-Si:H, and the estimation here from the photocurrent spectroscopy gives almost the same value after prolonged illumination. Note that at the initial stage of the illumination the QE is much higher than that after prolonged illumination for both a-As_2Se_3 and a-Si:H.

5.4 *In situ* simultaneous measurement of PVE, PD, and PDC: the dynamics of time evolution and mechanisms

Important issues concerning photoinduced effects on amorphous chalcogenides are still a matter of debate: is there any direct correlation among PVE, PD, and PDC? To formulate a proper answer to this question, *in situ* and simultaneous measurements of PVE, PD, and PDC have been performed in a-As_2Se_3 films. In the following, we concentrate on the difference in the dynamics of these photoinduced changes. Note that in previous sections we discussed PVE, PD, and PDC *independently* of each other.

Figure 5.16 shows the *in situ* simultaneous measurement system which monitors the changes of the surface height (for PVE), optical transmittance (for PD), and photocurrent (for PDC). A phase-shifting interferometer (PSI) was

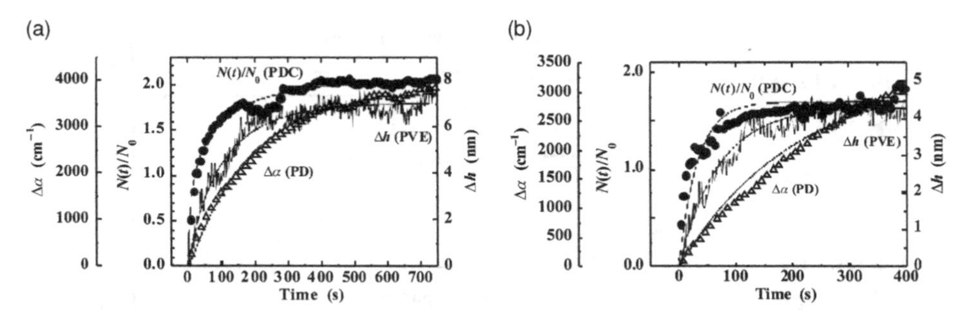

Figure 5.17. Time evolution of the change in $N(t)/N_0$, $\Delta\alpha(t)$, and surface height $\Delta h(t)$ for thicknesses of (a) $d = 700$ nm and (b) $d = 200$ nm in a-As_2Se_3. (From Nakagawa *et al.* (2010). *Phys. Status Solidi C*, **7**, 857. Copyright 2013 by John Wiley and Sons Inc.)

employed to measure the height map, as already described in Section 5.1. A He–Ne laser (633 nm, 200 mW cm^{-2}) was used as the exciting light source, and a DPSS laser (475 nm) was used to monitor the surface height. For PD, the transmittance of the He–Ne laser light passing through the sample was used; for PVC, the photocurrent induced by the He–Ne exciting light was used. The measurements were performed at 300 K in air (Nakagawa *et al.*, 2010).

Figures 5.17(a) and (b) show the time evolution of the changes in $N(t)/N_0$, $\Delta\alpha(t)$, and the surface height $\Delta h(t)$ for thicknesses $d = 700$ nm and $d = 200$ nm in a-As_2Se_3. Note that the optical penetration depth for a-As_2Se_3 is $\alpha^{-1} =$ 600 nm at $\lambda = 633$ nm, and hence 200 nm is significantly lower than α^{-1}. This means that for $d = 200$ nm, the illumination is almost homogeneous throughout the sample.

All changes are empirically represented by the following function:

$$y = A\left[1 - \exp\left\{-(t/\tau)^{\beta}\right\}\right], \tag{5.15}$$

where A is a constant (equal to the total change), and τ and β have the same meanings as described following eqn. (5.13). The fitting of eqn. (5.15) to the experimental data (thin solid curves in Figures 5.17(a) and (b)) produces the important physical parameters τ and β, which are indicated in Table 5.1.

In eqn. (5.15), the function with $\beta < 1.0$ is called the stretched exponential function (SEF). As indicated in Table 5.1, the dynamics of PDC ($N(t)/N_0$) are clearly represented by a SEF, although the origin of the SEF is not clear. On the other hand, the dynamics of PVE (Δh) are given by the exponential function ($\beta = 1.0$). However, note that for the dynamics of PD ($\Delta\alpha$) when $d = 700$ nm, β is 0.95, which is a lower value than that for $d = 200$ nm. We know that PDC and PVE always follow stretched exponential and exponential dynamics,

Table 5.1. *Physical parameters τ and β obtained for $N(t)/N_0$, $\Delta\alpha$, and Δh for each film thickness.*

	$d = 200$ nm		$d = 700$ nm	
	τ(s)	β	τ(s)	β
$N(t)/N_0$	18	0.55	40	0.55
Δh	70	1.0	90	1.0
$\Delta\alpha$	200	1.0	180	0.95

From Shimakawa, Nakagawa, and Itoh (2009).

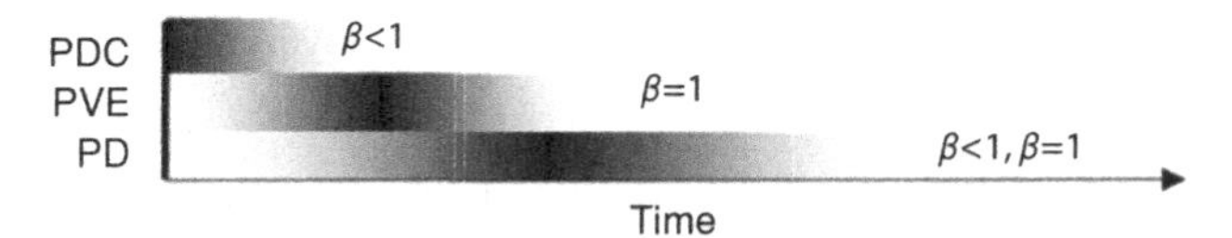

Figure 5.18. Comparison of the time scales in the evolutions of PDC, PVE, and PD, together with the parameter β. (From Nakagawa *et al.* (2010). *Phys. Status Solidi C*, **7**, 857. Copyright 2013 by John Wiley and Sons Inc.)

respectively, whereas the PD dynamics are expected to be dependent on the film thickness. The τ value is lowest for PDC and is highest for PD compared with the τ value for each phenomenon. These features are summarized in Figure 5.18. It is suggested that there is no direct one-to-one correlation among PDC, PVE, and PD. Thickness-dependent values of β and τ for PD and the origin of SFE for PD will be discussed later in this section using simulations.

Let us discuss the models that account for the experimental results so far observed. There are principally three models to account for the occurrence of PD: (1) the change in atomic positions (Tanaka, 1983), (2) bond breaking and/or bond alternation (Elliott, 1986; Kolobov *et al.*, 1997; Hegedus *et al.*, 2005), and (3) the expansion and slip motions of charged layers (Shimakawa *et al.*, 1998). This last model can be used to interpret current results of *in situ* simultaneous measurements in a unified way, and we discuss this now.

As shown in Figure 5.19, clustered layers, for example As_2Se_3, become negatively charged (the holes diffuse away to the unilluminated region), and hence a repulsive interlayer Coulombic interaction produces an increase in the interlayer distance (volume expansion). This process is indicated by the "Expansion" arrow. A slip motion along the layers, indicated by the "Slip" arrow, should also take place following the expansion motion. This slip motion is known to *cause*

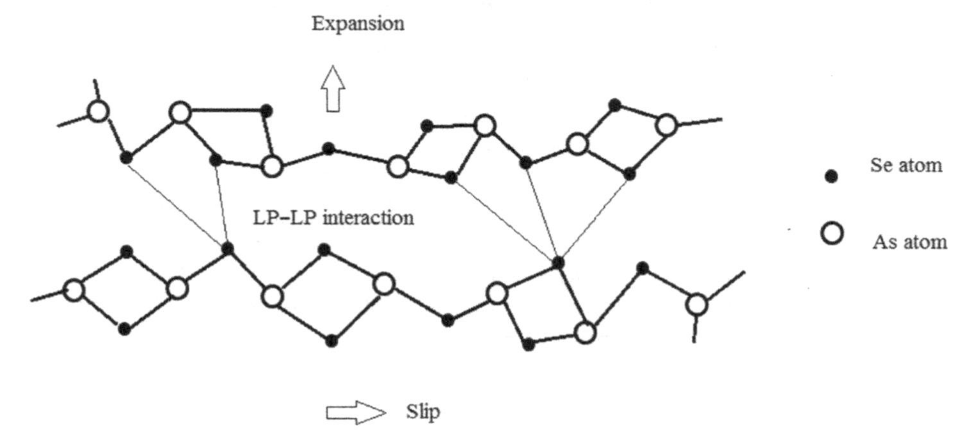

Figure 5.19. Schematic of photoinduced effects in layered chalcogenides.

an increase in lone-pair interactions between interlayers, leading to PD. The PDC corresponds to bond breaking, which occurs principally in intralayers (for example, the bond breaking between As and Se in Figure 5.19), which is discussed in Section 5.3. The intralayer bond breaking may not have a significant role for PD or PVE. However, there are some suggestions that a correlation between PDC and PD exists (Elliott, 1986; Kolobov *et al.*, 1997; Hegedus *et al.*, 2005).

Finally, it is of interest to discuss the dynamics of PD, i.e. the time evolution of PD, since the dynamic function is given by an exponential function ($\beta = 1.0$) in thin films that changes to a stretched exponential function ($\beta < 1.0$) in thicker films. As PD is observed with a change in the optical absorption coefficient (a change in optical transmittance along the thickness direction), the dynamics of PD may depend on the thicknesses of the films. If the dynamics changed with thickness, we would be able to identify the origin of the stretched exponential function in the PD occurrence and discuss the correlation between PD, PVE, and PDC. Note that the stretched exponential dynamics for numerous phenomena in disordered matters has been a perplexing issue for many decades (Palmer *et al.*, 1984; Shimakawa, 1984, 1985, 1986; Phillips, 1996).

A model calculation for β and τ has been made in a-As_2Se_3 by assuming a *series sequence* of PD along the thickness direction from the illuminated surface to the back surface (Shimakawa *et al.*, 2009). We know that the optical absorption decreases exponentially with the thickness direction toward the non-illuminated back surface. It is assumed that the reaction rate (the inverse of the response

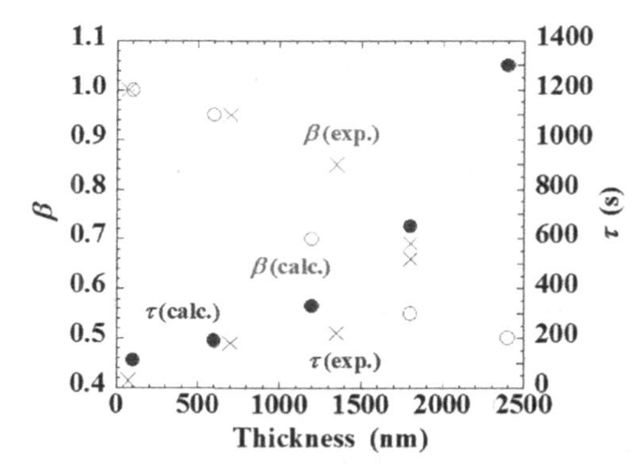

Figure 5.20. Variations from the model calculation (denoted by ×) in β and τ, together with the experimental results (denoted by circles) in a-As_2Se_3 films. (Taken with permission from Shimakawa *et al.* (2009). *Appl. Phys. Lett.*, **95**, 051908. Copyright 2013, American Institute of Physics.)

time) for a PD occurrence is proportional to the number of absorbed photons in each layer. It is further assumed that in each layer the PD dynamics is dominated by an exponential function.

Figure 5.20 shows variations from the model calculation (denoted by ×) in β and τ, together with the experimental results (denoted by circles) in a-As_2Se_3 films. We can see that β and τ obtained from the simulation decreased and increased with the growing film thickness, respectively, and that they replicate the experimental results well. It is thus concluded that the dynamic response of the PD itself essentially follows the exponential function (non-dispersive), and the stretched exponential dynamics occur only in thicker films, due to a series sequence of PD occurrences layer by layer. Thus, a stretched dynamic function for PD is not an essential feature here. Instead, as already discussed in Section 5.3, the dynamics represented by the stretched exponential function is essential to the nature of PDC processes. The origin of the stretched exponential function found only in PDC has been discussed recently in the literature (Freitas, Shimakawa, and Wagner, 2014), in which a special distribution of the reaction rate in PDC is assumed.

5.5 Photocrystallization and photoinduced amorphization

Unlike the *reversible* photoinduced changes, photoinduced or photon-assisted crystallization is an *irreversible* change. Crystallization sometimes leads to the damage of optical devices using amorphous states; therefore, we will discuss it briefly first in this section. We then return to the *reversible* photoamorphization observed in some materials. The term "photon assisted" can be used for the condition when the photon irradiation *enhances* some effects (for example, crystallization using thermal energy). Here we use the term *photocrystallization* as a photon effect.

In a-Se films, the crystal growth rate increases with the illumination, and the growth rate is controlled by the *hole* flux toward amorphous–crystalline boundaries (Dresner and Stringfellow, 1968). This type of photocrystallization may be related to the suggestion (in Section 5.1) that holes expand an a-Se network (Hegedus *et al.*, 2005). Inami and Adachi (1999) discovered that oriented crystallization occurs upon linearly polarized illumination. Photocrystallization has also been reported in a-$GeSe_2$ (Sakai *et al.*, 2003), As_xSe_{1-x} (Mikla and Mikhalko, 1995), and AsS (Brazhkin *et al.*, 2007).

A crystallized $As_{50}Se_{50}$ film on silica glass substrate becomes amorphous again when continuous-wave low-intensity light is irradiated. This change is athermal and is not caused by local melting (and the subsequent quenching) (Elliott and Kolobov, 1991; Kolobov and Elliott, 1995). The x-ray diffraction (XRD) patterns of a thermally crystallized and illuminated film clearly show the difference, and XRD patterns vary markedly with the film thickness, which implies that the substrate plays a role in the amorphization process. A realgar-like As_4Se_4 molecular structure of the crystallized film is confirmed by the Raman spectrum. Annealing at a temperature above the glass-transition temperature results in a different XRD pattern (not the same as that for the originally crystallized film). It should be noted that this phenomenon seems to be unique to this composition ($As_{50}Se_{50}$) (Elliott and Kolobov, 1991).

5.6 Some applications of photoinduced effects

Photoinduced effects are enhanced in the presence of localized plasmon fields, which can be caused by an increase in the non-equilibrium charge carrier generation in semiconductors. This has been observed in a-Ch–gold nanoparticle (GNP) composite structures (Charnovych and Kökényesi, 2011; Charnovych *et al.*, 2011; Voynarovych *et al.*, 2012). Note also that the localized plasmon fields enhance the efficiency, for example the performance of solar cells, as well as the photoinduced structural transformations in amorphous chalcogenides. The transformation of the GNP plasmon resonance spectrum due to the changes of complex optical parameters of the surrounding chalcogenide matrix was used for a more profound understanding of the photoinduced process in the given material (Charnovych and Kökényesi, 2011; Charnovych *et al.*, 2011; Voynarovych *et al.*, 2012).

The photoinduced volume change within the amorphous phase consists of two components: a rapid but minor one connected with a local structural change, and a slow one related to a photostimulated mass transport (Kaganovskii *et al.*, 2011; Trunov *et al.*, 2011), which can also be influenced by plasmon fields (Charnovych *et al.*, 2013).

The photostimulated mass transport is also a basic effect in chalcogenide/chalcogenide or metal/chalcogenide nanomultilayers, which results in an essentially enhanced intermixing of neighboring nanometer-thick layers (for example, a-Se/As_2S_3, GeS/GeSe, etc.) (Kökényesi, 2006; Naik *et al.*, 2009; Takats *et al.*, 2009; Charnovych *et al.*, 2013) and provides an exceptionally interesting basis for direct surface pattern recording. Further development of photoresist or even electron-beam resist materials based on a one-step recording process (Cserhati *et al.*, 2012) for the prototype development of optoelectronic elements, as well as new optical non-linear media, may be expected from these recent results.

References

Branz, H.M. (1999). Hydrogen collision model: quantitative description of metastability in amorphous silicon. *Phys. Rev. B*, **59**, 5498–5512.

Brazhkin, V.V., Gavrilyuk, A.G., Lyapin, A.G., Timofeev, Yu.A., Katayama, Y., and Kohara, S. (2007). AsS: bulk inorganic molecular-based chalcogenide glasses. *Appl. Phys. Lett.*, **91**, 031912, 1–3.

Charnovych, S. and Kökényesi, S. (2011). Plasmon enhanced optical recording in As–Se layers. *Phys. Status Solidi C*, **8**, 2854–2857.

Charnovych, S., Kökényesi S., Glodan, Gy., and Csik, A. (2011). Enhancement of photoinduced transformations in amorphous chalcogenide film via surface plasmon resonances. *Thin Solid Films*, **519**, 4309–4312.

Charnovych, S., Szabó, I.A., Tóth, A., Volk, J., Trunov, M.L., and Kökényesi, S. (2013). Plasmon assisted photoinduced surface changes in amorphous chalcogenide layer. *J. Non-377Cryst. Solids*, 200–204.

Cserhati, C., Charnovych, S., Lytvyn, P.M. *et al.* (2012). Mechanism of photo induced mass transfer in amorphous chalcogenide films. *Mater. Lett.*, **66**, 159–161.

Dresner, J. and Stringfellow, G.B. (1968). Electronic processes in the photo-crystallization of vitreous selenium. *J. Phys. Chem. Solids*, **29**, 303–311.

Elliott, S.R. (1986). A unified model for reversible photostructural effects in chalcogenide glasses. *J. Non-Cryst. Solids*, **81**, 71–98.

Elliott, S.R. and Kolobov, A.V. (1991). Athermal light-induced vitrification of $As_{50}Se_{50}$ films. *J. Non-Cryst. Solids*, **128**, 216–220.

Freitas, R.J., Shimakawa, K., and Wagner, T. (2014). The dynamics of photoinduced defect creation in amorphous chalcogenides: the origin of the stretched exponential function. *J. Appl. Phys.*, **115**, 013704, 1–4.

Fritzshe, H. (2000). Light-induced effects in glasses. In *Insulating and Semiconducting Glasses*, ed. P. Boolchand. Singapore: World Scientific, pp. 653–690.

Ganjoo, A. and Shimakawa, K. (2002). Dynamics of photodarkening in amorphous chalcogenides. *J. Optoelectron. Adv. Mater.*, **4**, 595–604.

Ganjoo, A., Ikeda, Y., and Shimakawa, K. (1999). *In situ* photoexpansion measurements of amorphous As_2S_3 films: role of photocarriers. *Appl. Phys. Lett.*, **74**, 2119–2122.

Ganjoo, A., Shimakawa, K., Kamiya, H., Davis, E.A., and Singh, J. (2000). Percolative growth of photodarkening in amorphous As_2S_3 films. *Phys. Rev. B*, **62**, R14601–14604.

Ganjoo, A., Shimakawa, K., Kitano, K., and Davis, E.A. (2002). Transient photodarkening in amorphous chalcogenides. *J. Non-Cryst. Solids*, **299–302**, 917–923.

Gotoh, T., Nonomura, S., Nishio, M., Nitta, S., Kondo, M., and Matsuda, A. (1998). Experimental evidence of photoinduced expansion in hydrogenated amorphous silicon using bending detected optical lever method. *Appl. Phys. Lett.*, **72**, 2978–2980.

Hamanaka, H., Tanaka, K., Matsuda, A., and Iijima, S. (1976). Reversible photo-induced changes in evaporated As_2S_3 and $As_4Se_5Ge_1$ films. *Solid State Commun.*, **19**, 499–501.

Hegedus, J., Kohary, K., Pettifor, D.G., Shimakawa, K., and Kugler, S. (2005). Photo-induced volume changes in amorphous selenium. *Phys. Rev. Lett.*, **95**, 206803, 1–4.

Hegedus, J., Kohary, K., and Kugler, S. (2006). Universal feature of photo-induced volume changes in chalcogenide glasses. *J. Non-Cryst. Solids*, **352**, 1587–1590.

Ikeda, Y. and Shimakawa, K. (2004). Real-time in situ measurements of photoinduced volume changes in chalcogenide glasses. *J. Non-Cryst. Solids*, **338**–340, 539–542.

Inami, T. and Adachi, S. (1999). Structural and optical properties of photocrystallized Se films. *Phys. Rev. B*, **60**, 8284–8289.

Itoh, N. and Stoneham, A.M. (2001). *Materials Modification by Electronic Excitation*. Cambridge: Cambridge University Press.

Kaganovskii, Yu., Beke, D.L., Charnovych, S., Kökényesi, S., and Trunov, M.L. (2011) Inversion of the direction of photo-induced mass transport in $As_{20}Se_{80}$ films: experiment and theory. *J. Appl. Phys.*, **110**, 063502, 1–5.

Kökényesi, S. (2006). Amorphous chalcogenide nano-multilayers: research and development. *J. Optoelectron. Adv. Mater.*, **8**, 2093–2096.

Kolobov, A.V. (ed.) (2003). *Photo-Induced Metastability in Amorphous Semiconductors*. Weinheim: Wiley-VCH.

Kolobov, A.V. and Elliott, S.R. (1995). Reversible photo-amorphization of a crystallized $As_{50}Se_{50}$ alloy. *Philos. Mag. B*, **71**, 1–10.

Kolobov, A.V., Oyanagi, H., Tanaka, K., and Tanaka, K. (1997). Structural study of amorphous selenium by *in-situ* EXAFS: observation of photoinduced bond alternation. *Phys. Rev. B*, **55**, 726–734.

Kugler, S., Hegedus, J., and Kohary, K. (2007). Modelling of photoinduced changes in chalcogenide glasses: a-Se and a-As_2Se_3. *J. Mater. Sci.: Mater. Electron.*, **18**, S163–S167.

Kuzukawa, Y., Ganjoo, A., Shimakawa, K., and Ikeda, Y. (1999). Photo-induced structural changes in obliquely deposited arsenic-based amorphous chalcogenides: a model for photostructural changes. *Philos. Mag. B*, **79**, 249–256.

Lukács, R. and Kugler, S. (2011). Photoinduced volume changes of obliquely and flatly deposited amorphous AsSe films: universal description of the kinetics. *Jpn. J. Appl. Phys.*, **50**, 091401, 1–4.

Lukács, R., Hegedus, S., and Kugler, S. (2009). Microscopic and macroscopic models of photo-induced volume changes in amorphous selenium. *J. Mater. Sci.: Mater. Electron.*, **20**, S33–S37.

Lukács, R., Veres, M., Shimakawa, K., and Kugler, S. (2010). On photoinduced volume change in amorphous selenium: quantum chemical calculation and Raman spectroscopy. *J. Appl. Phys.*, **107**, 073517, 1–5.

Mikla, V.I. and Mikhalko, I.P. (1995). Laser-induced structural transformation of As_xSe_{1-x} thin amorphous films. *J. Non-Cryst. Solids*, **180**, 236–243.

Molina, D., Lomba, E., and Kahl, G. (1999). Tight-binding model of selenium disordered phases. *Phys. Rev. B*, **60**, 6372–6382.

Morigaki, K. (1999). *Physics of Amorphous Semiconductors*. London: World Scientific & Imperial College Press.

Morigaki, K. and Hikita, H. (2007). Modeling of light-induced defect creation in hydrogenated amorphous silicon. *Phys. Rev. B*, **76**, 085201, 1–17.

Mott, N.F. and Davis, E.A. (1979). *Electronic Processes in Non-Crystalline Materials*, 2nd edn. Oxford: Oxford University Press.

Naik, R., Adarsh, K.V., Ganesan, R. *et al.* (2009). X-ray photoelectron spectroscopic studies on Se/As_2S_3 and Sb/As_2S_3 nanomultilayered film. *J. Non-Cryst. Solids*, **355**, 1836–1839.

Nakagawa, N., Shimakawa, K., Itoh, T., and Ikeda, Y. (2010). Dynamics of principal photoinduced effects in amorphous chalcogenides: in situ simultaneous measurements of photodarkening, volume changes, and defect creation. *Phys. Status Solidi C*, **7**, 857–860.

Palmer, R.G., Stein, D.L., Abraham, E., and Anderson, P.W. (1984). Models of hierarchically constrained dynamics for glassy relaxation. *Phys. Rev. Lett.*, **53**, 958–961.

Pfeifer, G., Paesler, M.A., and Agarwal, S.C. (1991). Reversible photodarkening of amorphous arsenic chalcogens. *J. Non-Cryst. Solids*, **130**, 111–143.

Phillips, J.C. (1996). Stretched exponential relaxation in molecular and electronic glasses. *Rep. Prog. Phys.*, **59**, 1133–1207.

Redfield, D. and Bube, R.H. (1996). *Photoinduced Defects in Semiconductors*. New York: Cambridge University Press.

Sakai, K., Maeda, K., Yokoyama, H., and Ikari, T. (2003). Photo-enhanced crystallization by laser irradiation and thermal annealing in amorphous $GeSe_2$. *J. Non-Cryst. Solids*, **320**, 223–230.

Shimakawa, K. (1984). Origin of nonsymmetric dielectric relaxation in dipolar materials. *Appl. Phys. Lett.*, **45**, 587–588.

Shimakawa, K. (1985). Exciton recombination in amorphous chalcogenides. *Phys. Rev. B*, **31**, 4012–4014.

Shimakawa, K. (1986). Persistent photocurrent in amorphous chalcogenides. *Phys. Rev. B*, **34**, 8703–8708.

Shimakawa, K., Inami, S., and Elliott, S.R. (1990). Reversible photoinduced change of photoconductivity in amorphous chalcogenide films. *Phys. Rev. B*, **42**, 11857–11861.

Shimakawa, K., Inami, S., Kato, T., and Elliott, S.R. (1992). Origin of photoinduced metastable defects in amorphous chalcogenides. *Phys. Rev. B*, **46**, 10062–10069.

Shimakawa, K., Kolobov, A.V., and Elliott, S.R. (1995). Photoinduced effects and metastability in amorphous semiconductors and insulators. *Adv. Phys.*, **44**, 475–588.

Shimakawa, K., Yoshida, N., Ganjoo, A., Kuzukawa, Y., and Singh, J. (1998). A model for the photostructural changes in amorphous chalcogenides. *Philos. Mag. Lett.*, **77**, 153–158.

Shimakawa, K., Ikeda, Y., and Kugler, S. (2004a). Fundamental optoelectronic processes in amorphous chalcogenides. In *Optoelectronic Materials and Devices, Vol.1*, eds. G. Lucovsky and M. Popescu. Bucharest: INOE Publishing House, p. 103.

Shimakawa, K., Mehern-Nessa, Ishida, H., and Ganjoo, A. (2004b). Quantum efficiency of light-induced defect creation in hydrogenated amorphous silicon and amorphous As_2Se_3. *Philos. Mag. B*, **84**, 81–89.

Shimakawa, K., Nakagawa, N., and Itoh, T. (2009). The origin of stretched exponential function in dynamic response of photodarkening in amorphous chalcogenides. *Appl. Phys. Lett.*, **95**, 051908, 1–3.

Shluger, A. and Stefanovich, E. (1990). Models of the self-trapped exciton and nearest-neighbor defect pair in SiO_2. *Phys. Rev. B*, **42**, 9664–9673.

Singh, J. and Shimakawa, K. (2003). *Advances in Amorphous Semiconductors*. New York and London: Taylor and Francis.

Song, K.S. and Williams, R.T. (1996). *Self-Trapped Excitons*, 2nd edn. Berlin: Springer.

Staebler, D.L. and Wronski, C.R. (1977). Reversible conductivity changes in discharge-produced amorphous Si. *Appl. Phys. Lett.*, **31**, 292–294.

Street, R.A. (1991). *Hydrogenated Amorphous Silicon*. Cambridge: Cambridge University Press.

Takats, V., Miller, F., Jain, H., Cserhati, C., and Kökényesi, S. (2009). Direct surface patterning of homogeneous and nanostructured chalcogenide layers. *Phys. Status Solidi C*, **6**(suppl. 1), S83–S85.

Tanaka, K. (1983). Mechanisms of photodarkening in amorphous chalcogenides. *J. Non-Cryst. Solids*, **59–60**, 925–928.

Tanaka, K. (1990). Photoinduced structural changes in chalcogenide glasses. *Rev. Solid State Sci.*, **4**, 641–659.

Tanaka, K. (1998). Photoexpansion in As_2S_3 glass. *Phys. Rev. B*, **57**, 5163–5167.

Tanaka, K. and Shimakawa, K. (2011). *Amorphous Chalcogenide Semiconductors and Related Materials*. New York: Springer.

Trunov, M.L., Lytvyn, P.M., Yannopoulos, S.N., Szabo, I.A., and Kökényesi, S. (2011). Photoinduced mass-transport based holographic recording of surface relief gratings in amorphous selenium films. *Appl. Phys. Lett.*, **99**, 051906, 1–3.

Uchino, T., Takahashi, T., Ichii, M., and Yoko, T. (2002). Microscopic model of photoinduced and pressure induced UV spectra changes in germanosilicate glass. *Phys. Rev. B*, **65**, 172202, 1–4.

Voynarovych, I., Kökényesi, S., Yurkovich, N., Charnovych, S., and Dmitruk, N. (2012). Plasmon-assisted transformations in metal-amorphous chalcogenide light-sensitive nanostructures. *Plasmonics*, **7**, 341–345.

Index

Printed in the United States
By Bookmasters